Herbal and Acidulant Tree Spices

Kodoth Prabhakaran Nair

Herbal and Acidulant Tree Spices

Their Global Commercial Potential

Springer

Kodoth Prabhakaran Nair
Villament G3
C/o Dr. Mavila Pankajakshy
Kozhikode, Kerala, India

ISBN 978-3-031-04359-8 ISBN 978-3-031-04360-4 (eBook)
https://doi.org/10.1007/978-3-031-04360-4

This Springer imprint is published by the registered company Springer Nature Switzerland AG
The registered company address is: Gewerbestrasse 11, 6330 Cham, Switzerland

I did dedicate this book to my family: my wife, Dr. Pankajam Nair, a nematologist trained in Europe, who gave up her profession to be a homemaker, more than four decades ago, when we had our son and daughter, who is my all, and one who sustains me in this difficult journey that life is; our son, Dr. Kannan, who assisted me in gathering a lot of scientific literature; and our daughter Engineer Sridevi and her husband Engineer Arvind, who constantly encouraged me to write this book against all the external constraints imposed by the Covid-19 pandemic.

Acknowledgements

A word of true appreciation:

Ms. Margaret Deignan, senior editor and publisher, has extended unwavering support to my writing assignments, and I owe her a deep debt of gratitude.

Ms. Deepthi Vasudevan, a very conscientious book editor, who guided me diligently through the production process, and Ms. Rekhajaykumar and her production team who did a marvelous job in fast tracking production of this important book, which I hope will be enthusiastically accepted by a global readership.

Contents

Introduction

Abstract The chapter discusses, at length, the great economic potential on a global scale of the world's unique herbal and acidulant tree spices.

Keywords Herbal spice · Acidulant tree spice · International trade

Leafy herbal spices are well known as "Aromatic herbs", which, apart from being used in the culinary art of flavoring of foods and beverages , are also known to possess several utility properties, such as, antiseptic, anti-oxidant, pharmaceutical and nutraceutical properties. Of the several leafy herbal spices, only the following sixteen are discussed in this book. It is expected that this comprehensive book will provide a platform for tapping their research potential and global trade potential.

Economic Importance of Herbal Spices: International Trade

According to the International Trade Forums survey of January-March 1993, an idea of their competitive economics in international trade can be visualized from the following table (Table 1), which shows that 12 commercially important herbal spices are imported into 4 major European markets to the tune of 12,000–13,000 tonnes, per annum and there is much scope for their greater demand.

France is the biggest importer of these herbs. According to trade estimates, the overall French demand for dried culinary herbs, both domestically produced and imported amount to about 6000–7000 tonnes per annum of which 75% is through imports. Likewise, the United Kingdom demand for herbs now runs to about 4000 tonnes per annum of which 65% is imported. Germany is the second largest consumer of culinary herbs in Europe, next to France; German demand is estimated to be about 5500 tonnes per annum of which 71% is imported. The Netherlands, though a small consumer (about 1500 tonnes per annum) is the largest re-exporter of culinary herbs to the tune of about 1500 tonnes per annum, through its major port

K. P. Nair, *Herbal and Acidulant Tree Spices*,
https://doi.org/10.1007/978-3-031-04360-4_1

Rotterdam, the largest transshipment port for spices and herbs. Interestingly, the Dutch has a great legacy in spice trade preceding that of the British.

Thankamani et al. (1994) have published an excellent exhaustive review on important agronomic aspects of four aromatic tree spice crops, namely, cinnamon, clove, nutmeg and pimento (allspice). The summarized available general information is given in the Table 2. The spices/condiments which are used only to acidify curries, fall into this group, and, in brief are described below in the following table (Table 3).

Table 1 Import of dried herbs to 4 major European markets (in tonnes)

Herb	France	Germany	United Kingdom	The Netherlands	Total
Basil	300–350	200	250	80	330–880
Bay leaves	250–300	250	185	70	755–805
Dill	20	20	30	25	95
Marjoram	530	500	252	400	1682
Mint	1000–1500	400	400	50	1850–2350
Oregano	350–500	600	500	150	1750
Parsley	100–200	50	100	400	650–750
Rosemary	350–400	450	90	80	970–1020
Sage	450	650	500	150	1750
Savory	120	100	15	30	265
Tarragon	50	65	13	25	153
Thyme	700–770	500	220	90	1510–1580

Source: International Trade Forums- March (1993)

Table 2 General information on world's important aromatic tree spices

Name of tree spice	Economic part used	Origin	Major producing countries	India		
				Area(ha)	Production (tonnes)	Major growing states
Clove	Dried unopened	Moluccas	Indonesia	1735	1477	Kerala
Eugenia caryophyllussyn	flower buds	Islands of Indonesia	Tanzania, Sri Lanka			Tamil Nadu
Syzygium aromaticum (Linn.) Merril& Perry Family: Myrtaceae			Madagascar Andaman and Nicobar			
Nutmeg (Mace)	Seed kernel	Moluccas	Indonesia	38,238	3663	Kerala
Myristica fragrans			Islands of Indonesia			
Houtt						
Family:Myrtaceae			West Indies, Grenada			Tamil Nadu

(continued)

Table 2 (continued)

Name of tree spice	Economic part used	Origin	Major producing countries	India		
				Area(ha)	Production (tonnes)	Major growing states
Cinnamon/Cassia	Bark	Sri Lanka	Indonesia	565	355	Kerala & Tamil Nadu
Cinnamomum verum		Malabar	Tanzania			
Bercht&Presl.						
Family:Lauraceae		Coast of India	Sri Lanka Seychelles, Madagascar			
Allspice/Pimento	Dried fruit Berries	West Indies	Jamaica	N.A.	N.A.	Kerala and Tamil Nadu
Pimenta dioica Linn.						
Family:Myrtaceae		Central America	Guatemala, Mexico, Honduras, Cosa Rica			

Source: Thankamani et al. (1994)
N.A. not Available

Table 3 General information of four important acidulant tree spices

S.No.	English name	Botanical name	Family	Plant part used
1.	*Amchur*	*Mangifera indica*	Anacardiaceae	Dried flesh of peeled green mangoes (mature but unripe)
2.	*Anardana*	*Punica granatum*	Punicaceae	Dried seed with flesh of pomegranate
3.	Kokum or kokam	*Garcinia indica*	Guttiferae	Fruit rind
4.	Tamarind	*Tamarindus indica*	Leguminosae	Ripe fruit (pod) Pulp

Reference

Thankamani CK, Sivaraman K, Kandiannan K, Peter KV (1994) Agronomy of tree spices (Clove, Nutmeg, Cinnamon and Allspice or Pimento). A review. J Spices Aromat Crops 3(2):105–123

Anardana

Abstract The chapter will cover various aspects, such as, description, distribution, area, production and export, crop management, with special emphasis on soil, climate, flowering, fruiting and after care. In addition, it would also discuss details of several varieties of Anardana, plant protection, quality standards, and processed products

Keywords Description · Distribution · Flowering · Varieties · Quality standards

Botanical name: *Punica granatum* Linn.
Family: Punicaceae

Description and Distribution

Anardana comprises dried seeds (with flesh or pulp on) of pomegranate which is a shrub or a small tree, about 5–8 m tall. It is considered a native of Afghanistan and Baluchistan.

It is found growing wild in warm valley and the outer hills of the Himalayas, between 900 and 1800 m. It is also cultivated at select sites almost throughout India. The pomegranate is an ancient and popular fruit. It is symbolic of plenty and prosperity. It grows best in the tropics, below an elevation of 100 m, with long, hot, dry summer and cool winters, or in those areas which are continuously warm and dry, provided, irrigation is available. High temperature should accompany the ripening season. The fruit is a large, globose berry, shiny red, tan, brown, yellowish-green or whitish in color, when ripe, crowned by the calyx, and is generally 5–7.5 cm in diameter. The fruit is filled with angular hard seeds which are covered with juicy red, pink or yellowish white, sweet-astringent acid pulp. It is the seeds with pulp on, which can be sundried or dehydrated, which constitute the condiment, *Anardana*.

K. P. Nair, *Herbal and Acidulant Tree Spices*,
https://doi.org/10.1007/978-3-031-04360-4_3

Like *Amchur*, *Anardana* also is used as an acidulant in Indian cooking, mostly in Northern States of India.

A wild type of pomegranate called "*Daru*" grows abundantly in lower Himalayas. Its seeds, when dried, give good quality *Anardana*. Another source of *Anardana* is said to be the wild pomegranate trees covering vast strips of hill slopes of Jammu and parts of Chamba, Kangra and Mandi districts of Himachal Pradesh and Punjab state, in India. The main areas from where the *Anardana* is collected are the "Raisu", Udhampur, Ramban, Kishtwar and Bhadurwah Forest Divisions. In these areas, the fruits are hand-picked towards the middle of October, when they are ripe and brownish-red in color. Seeds along with the pulp are separated by hand from the rind and are sun-dried for about 10–15 days when their color turns reddish-brown. On drying, it is called "*Anardana*".

Area, Production and Export

In India, pomegranate is grown on a small-scale in almost all the states. Maharashtra has 880 ha under pomegranate, Karnataka has 1320 ha, followed by Andhra Pradesh, Uttar Pradesh, Tamil Nadu, Gujarat, Rajasthan, and, Himachal Pradesh. Export of fresh pomegranates from India during 1992–1993 to 1994–1995 are given in the following Table 1. During 1994–1995 and 1996–1997, pomegranate worth INR worth 65.31 million and INR 66.93 million respectively were made.

The main center for forest *Anardana* is Udhampur, from where distribution is done all over the country. About 350 tonnes of *Anardana* is reported to have been marketed from Udhampur during 1963–1664. More recent statistics of production is not available. However, as far as fruit is concerned, pomegranate is grown in India in an area of 60,300 ha with an annual production of 3015 tonnes of fruit. Its production is mainly in Maharashtra State in India on a commercial scale (Rao and Purushotham 1988).

Exports of *Anardana*

The following Table 2 details the export data during 1996–1997.

Crop Management

Soil and Climate It grows all over India, but, thrives best in places with a hot dry summer, provided irrigation is available.

Table 1 Export of fresh pomegranate from India (quantity in kg) and (value in INR million)

	1992–1993		1993–1994		1994–1995	
Importing countries	Quantity	Value	Quantity	Value	Quantity	Value
UAE	542,766	7,145,098	1,303,709	17,924,574	1,737,852	26,544,234
Saudi Arabia	469,147	6,551,215	555,959	8,848,667	834,356	11,848,249
Bangladesh	238,167	2,332,502	3,372,213	3,539,936	1,019,583	8,975,691
Bahrain	99,977	1,335,289	228,472	3,486,103	192,409	3,037,269
Kuwait	52,901	729,377	81,446	1,263,203	141,655	1,893,769
UK	42,228	897,073	9133	196,868	76,915	1,380,735
South Africa	–	–	–	–	20,430	508,738
Oman	12,877	189,259	4724	70,835	14,874	403,768
Sri Lanka	–	–	15,555	251,656	19,752	297,069
Qatar	50,695	514,148	22,924	339,255	16,219	261,266
Pakistan	7578	127,725	–	–	37,474	217,355
Kenya	–	–	–	–	6779	167,106
Canada	3835	58,381	2560	54,418	11,899	92,088
Switzerland	5700	82,536	5015	110,099	4954	90,740
Netherlands	–	–	–	–	2200	81,687
Germany	17,405	671,952	66	875	3246	63,728
Singapore	6232	149,660	5990	151,971	2050	48.643
France	1368	41,632	20	520	555	14,265
Sweden	–	–	89	2086	1009	9790
Australia	–	–	–	–	77	4.667
USA	35	360	1112	20,221	200	3675
Hongkong	24,985	391,367	16,390	279,554	–	–
Tanzania	–	–	4000	130,891	–	–
Malaysia	2040	48,000	1000	34,000	–	–
Indonesia	–	–	850	10,000	–	–
Denmark	–	–	10	260		
Portugal	8	239	–	–	–	–
Mauritius	–	–	–	–	–	–
Brazil	–	–	–	–	–	–
Total product exported	1,627,944	2,265,823	2,623,237	36,715,992	4,144,488	55,944,532

In humid weather, the quality of fruit is not good. Pomegranate can withstand considerable frost and grows up to an altitude of 1600 m above sea level. When the winter is cold, the tree is deciduous, but, in the plains of India, it is ever green. It can tolerate alkaline and wet soils.

Propagation Many pomegranate plantations in India have been raised from seedlings as in the case of guava, but, the seedlings vary too much. Propagation from hard-wood cutting is easy. Layering is also possible. The planting distance is gener-

Table 2 Export of dried pomegranate seeds from India (quantity in kg and value in INR)

	1992–1993		1993–1994		1994–1995	
Importing countries	Quantity	Value	Quantity	Value	Quantity	Value
USA	1329	120,289	4526	321,483	3615	253,888
Canada	1557	164,566	900	63,962	2259	106,331
Sweden					184	44,342
New Zealand					575	44,310
Germany					240	19,544
UAE	–	–	150	9381	2201	1364
Kuwait					100	6723
Australia			290	29397	50	3752
Hongkong			43	6407	38	3546
Malaysia					25	1762
Singapore			10	1001	10	1112
Oman					50	938
Saudi Arabia			225	17,750	–	–
Botswana	–	–	25	2502		
UK	1600	153,680	–	–	–	–
Bahrain	–	–	–	–	–	–
Swaziland	–	–	–	–	–	
Total Export:	4486	438,455	6169	451,883	7366	497,632

Source: Daily lists of Customs House, Government of India, APEDA, Ministry of Commerce, Government of India

ally not adequate. Some plantations have trees as close as 3 meters. However, a distance of 6 meters from tree to tree is essential.

Flowering, Fruiting and After-Care

Pomegranate flowers in spring in North India, but, in central and South India, it flowers almost throughout the year. However, the flowers are most profuse early in the monsoon season. The spring flowering (*Ambe bahar*) gives fruits in summer, when the demand is maximum for the same. However, the fruit quality is poor. Flowering is, therefore, artificially induced early in the rainy season, as in the case of citrus. For this purpose, the mots are exposed in April and manuring and irrigation are done in May. After this treatment, the fruits ripen in November–December.

The fruits are borne terminally on short spurs arising from mature shoots. Shoots bear fruits in 3–4 years. Only a limited pruning of the bearing tree is, therefore, required. The gradual growth of new shoots should be encouraged by restricted cutting back of the bearing shoots. Suckers are removed regularly.

Varieties/Types of Pomegranate

In India, there are many types of pomegranate, including the double-flowering, ornamental type and a white-flowering type. The white-flowering type has a very poor fruit quality. In the cultivated types, the sugar content varies from 12% to 16% and the acidity from 1.5% to 2.5%.

Commercially Important Types/Varieties

Several types/varieties are cultivated in India. These differ in shape and size of the fruit, color and thickness of the rind, color of aril and mature seeds, degree of sweetness and/or acidity etc. Of the several types/varieties, the following are commercially important:

Alandi Fruits are medium in size, with blood-red or deep-pink arils and with sweet but slightly acidic pulp. Seeds are very hard. It is named after a village in Maharashtra from where it was first selected. It is popular in Maharashtra and North Karnataka.

Dholka Fruits are large and greenish-white colored; arils are fleshy, white or pinkish white, and sweet soft. It is a medium cropper. It is named after the place where it was cultivated on a large scale and spread to other places. It is a popular variety in Gujarat state, India.

Ganesh (GBG 1) It is a selection from seedling progenies made by late Dr. G.S.Cheema, the renowned Indian Horticulturist. It is an excellent, high yielding selection with fruits which are medium large to large, in size, arils are pink-colored, seeds are soft and high in total soluble solids (TSS). This is the most important commercial variety of Maharashtra state.

Bessein Seedless Fruits are medium-sized, with greenish-yellow skin. Arils are pinkish white. Seeds ae soft. The variety has been popular in Karnataka State.

Jyoth (GKVK) It is a selection from seedling progenies developed by the University of Agricultural sciences, Bengaluru, Karnataka State, India. It was released for commercial cultivation in 1985. The fruits are medium to large in size (200 g), have attractive yellowish color with deep-pink arils. Seeds are soft. It has high TSS and high pulp content (70.5%). This variety is becoming popular in Karnataka State and some parts of Maharashtra State.

Other Varieties The other varieties which have shown promise in South India (Kodur, Andhra Pradesh) are as follows:

Paper Shell Fruits are medium in size, rind is thick, aril is fleshy and deep pink colored. Seeds are soft, fruits are good flavored. It is a prolific bearer.

Spanish Ruby Fruits are small to medium in size, with soft seeds and thick skin. It is a prolific bearer.

Muskat Red Fruits are small to medium in size, rind is thick, arils are fleshy and sweet. Seeds are not very hard.

Other varieties of Importance in other countries:

Kabul Fruits are large, rind is thick and deep-red colored mixed with pale-yellow; arils are dark blood-red and juice slightly bitter

Kandhari Fruits are large, rind is deep-red; arils are fleshy and hard, blood-red in color, slightly acidic; but juice is sweet and seeds are hard.

Overseas varieties from Baghdad and Palestine: The following are some of the overseas varieties:

Selimi (from Baghdad) It has very big fruits weighing about 1 kg each, the fruit is juicy and sweet

Roman Chokab A black pomegranate (from Baghdad) bears fruit of dark color with large seeds. It has a tender sweet-sour taste.

Wellisai (from Palestine) is considered to be the best as it is soft-seeded and very early to bear.

Ras-el-Baghi (from Palestine) bears large, sweet and early maturing fruits.

Suffami (from Palestine) produces big juicy fruit with delicious sweet-sour taste.

The best specimen of this variety is obtained from Kabul, Afghanistan and Persia

Varieties from Spain

Dulca Colorado produces big fruits which are sweet and tasty

Grand Blanc produces seedless fruits

Commercial Varieties of California, USA

Choodesuy produces fruit of very high quality which stands transportation quite well

Boomageny produces large fruits, which are very juicy and stand shipment quite well

Classes or Types of Pomegranates from Cyprus

The pomegranates from Cyprus are grouped into three distinct classes as follows:

1. Sweet pomegranates locally known as *Glykia*
2. Mild sour pomegranates, neither sweet nor sour, but, of a mild sourish taste, locally known as Maifoshika
3. Sour pomegranate

Indian Pomegranates: Varieties and Types

Most of the commercially cultivated varieties fall under the following two groups:

1. The white-fleshed
2. The pink-fleshed

The quality and size of fruits, however, vary greatly, according to the locality where the plant is grown. The correct nomenclature and classification of the various indigenous types demand attention of the botanists and horticulturists of India, the earlier the better.

Overseas Varieties Sold in India

The following overseas types or varieties of pomegranate fruits are sold in Indian markets.

Muskati Fruit is of medium size, rind is russet green, seeds are soft, flesh is pinkish and sweet.

Kandhari Fruit is larger in size than Muskati; flesh is deep red, seeds are medium soft, color of the rind is deep red.

Kandhari Fruit is larger in size than Muskati; flesh is deep red, seeds are medium soft; color of the rind is deep red.

Sour Pomegranates They generally grow wild in the forests of the Himalayan mountains and are, as a rule, not sold in the fresh condition; but, they are available as a dried product called *Anardana* for use as a condiment in place of Tamarind, used in curries, such as Sambhar, as it is an acidulant.

Pomegranate is commercially grown in the mediterannian region, in countries such as Morocco, Egypt, Afghanistan and Baluchistan. Countries like China, the

USA, former USSR, Pakistan and India have also become production centers for this fruit.

Main Marketing Centers in India

Main marketing centers for pomegranate in India are Mumbai, in Maharashtra State, Pune and Sholapur also in Maharashtra State, Bengaluru in Karnataka State, Hyderabad in Andhra Pradesh, Delhi (Union territory) and Amritsar in Punjab State, and Jammu and Kashmir (also Union territories now). Marketing season varies from place to place. However, August to October is the peak period of marketing in many parts of India.

Economic Importance

Because of the hardiness of the plant and ease of cultivation, pomegranate is considered as an important cash crop on poor infertile soils of certain districts of Deccan in India. It does not tax the cultivator in any way by demanding much investment of monetary resources and labor. The income from the crop, though not huge, is fairly steady.

Limiting Factors

The extension of pomegranate cultivation is, however, not going on rapidly as far as one would wish, because the quality of the fruits grown under ordinary conditions of soil and climate are not very high. The fruit cracks badly as a result of variations in irrigation and other cultural methods, or possibly, due to the environment. These considerations limit the extension of the area under pomegranate. It is, however, to be hoped that extension will be favored by improved cultural methods, by evolving superior sour-and-sweet varieties and by devising suitable methods of processing.

Physiological Disorder

Splitting or Fruit Cracking The most serious problem of pomegranate cultivation in India is splitting or cracking of fruit. Sometimes more than 50% of fruits crack. It is due to hardening of skin of the fruits during periods of shortage of moisture and then sudden expansion of the inner part of the fruit after rains or heavy irrigation. Fruits are often harvested prematurely to avoid fruit-cracking, but, the resultant fruit quality is very poor in this case. Until now there is no cure for this problem. The problem can be minimized by improving the water-retention capacity of the soil; by applying organic manure to soil and by providing regular and light irrigation throughout the fruiting season. Some farmers apply powdered bricks to the soil in the belief that it would improve color of the fruit and the flesh. This is akin to the blind practice of applying milk to the roots of the mango tree in the belief that the mango fruit would be creamy and tasty. However, these practices probably, reduce fruit cracking by improving moisture-retention capacity of the soil.

Control of Fruit Cracking

Two foliar sprays of EPA (150ppm) and borfic acvid (1000ppm), the first after 15 days after full bloom and the second after 40 days before the expected harvest, were found to be very effective in controlling fruit-cracking of wild pomegranates at Palampur in Himachal Pradesh, India.

In Himachal Pradesh, premature fruit-cracking could be reduced when pomegranate was irrigated once a week or sprayed with 1% Potassium nitrate solution or 1% Magnesium sulfate solution or 0.025% Boric acid solution, after full bloom.

Plant Protection

Control of Insect Pests

Pomegranate Fruit Borer Among the different pests that attack the crop, the fruit borer (*Virachola isocrates*) is the most destructive which lowers the yield and quality of the fruit. The insect pest is also called "*Anar Butterfly*". Besides pomegranate, it is known to attack fruits of wood apple, guava, palm and tamarind.

The caterpillar damages the fruit. The full-grown caterpillar is short, stout, darkish brown in color with short hairs and whitish patches on the body. The caterpillar tunnels into the developing fruits and eats the seeds. And a rot sets in, bringing about the premature drop of the fruit. The fruit juice oozes out of the hole made by the borer. Thus, many of the fruits are destroyed.

The adult is a butterfly, medium in size, greyish-white in color with a wing span of 5 cm. It has a tail-like extension at the lower margins of lower wings. The female butterfly has conspicuous orange-colored patches on the fore-wings. The female butterfly lays whitish shiny eggs singly on flowers and tender fruits. They hatch into caterpillars within a week's time and enter the fruits by puncturing small holes and burrowing into the fruit.

The caterpillar becomes full grown in about a month's time, feeding on the pulp and seeds inside the fruit. The full-grown caterpillar makes a hole on the fruit and comes out for pupation on the stem or fruits fallen on the ground. And, the adult butterfly emerges after 7–34 days.

Control Measures The following preventive and control measures may be taken up in order to ensure satisfactory control of this insect pest. The following courses of treatment are recommended:

1. Wherever possible, young fruits may be enclosed loosely in punctured plastic bags (300 gauge) which are tied up at the neck of the bag. Butter paper or muslin cloth bags may be used for enclosing the fruits to prevent butterflies from laying eggs on the fruits.
2. The damaged fruits may be collected from the orchards and buried deep in the soil to prevent multiplication of the pest
3. Three sprays, starting from peak flowering stage, either with dimethoate (Rogor 30 EC-2 ml) or Endosulfan (Thiodan 35 EC -2 ml) or Carbaryl (4 g) or Phosphamidon (Demecron 85 EC- 1 ml) mixed with 1 liter of water, at 10 days interval, may be done on flowers and young fruits.

Once the caterpillar gets inside the fruit, no external treatment will help. Hence, for effective control of the insect pest spraying should be synchronized with the time when the eggs are laid on flowers or on the young fruits.

Spraying with calcium arsenate after fruit-set or dusting twice a month with a mixture of Paris green (4.07 kg), lime (2.26 kg), Talc (6.79 kg) and flour (113.4 g) has been recommended. DDT and BHC are not effective against the pest.

Other Caterpillars The caterpillar *Epijarbus ancus* which attacks fruits, has also been reported. Bark-eating caterpillar, *Aleurodes* sp., has also been found to attack the plant.

Chikta It is caused by small white insects *Aleurodes* sp. found in large numbers on the lower surface of the leaf. The insects secrete copious amounts of honey-dew on which black mold may develop. The damage is done by sucking the sap as well as by interfering in the photosynthesis which results in malformation of the fruits. The pest is controlled by spraying with Rosin compound prepared using the following formula:

Rosin	2.26 kg
Washing soda	0.45 kg
Water	22.73 liters

The Rosin should be powdered and mixed with washing soda in a vessel and add enough water. Boil the mixture and keep on adding a little of water until the mixture is clear and coffee-colored. When ripe, it should not give a milky appearance, on adding a drop of water to it. This stock solution should be further diluted for spraying in the ratio of 1:6.

Stem Borer Particularly, in old and neglected gardens the main trunks of the pomegranate trees are found badly infested with the stem borer caterpillars (*Arbella* sp.). Whenever the caterpillar is eating inside, a small hole will be seen and the network of the dried excreta of the insect will be found outside the hole. This has to be cleared and the insect should be extracted by inserting a fine-hooked wire into the hole. If this is not possible, a small cotton plug dipped in petrol, chloroform, or kerosene may be inserted and the whole should be closed with mud. The insect inside will die of suffocation caused by the vapors of the chemical. Considerable damage to plantations is caused by this stem borer. All the bark and other portions eaten by the insect should be cleaned and the damaged parts be treated with coal-tar to enable the wound to heal.

Squirrels Squirrels eat immature and mature fruits and cause serious losses. These can be kept away by baiting and enclosing half mature fruit (2–3 months old) in colored (blue color preferred) polythene bags. The bags should have vents made by cutting of the corners at the bottom.

Diseases

Fruit Rot This disease is caused by a fungus in pomegranate in Maharashtra state, India. The fruits rot badly, partially or wholly and become useless. The only remedy is to remove all the affected fruits and burn them. The fruits may be sprayed with Bordeaux mixture (5:5:50) as a preventive remedy.

Leaf Spot It is a serious disease causing leaf fall and spots on fruits. Spray thrice at fortnightly intervals with Dithane M-45 (0.2%) or Dithane Z-78 (2%) or Bavastin (0.50%).

Harvesting The trees begin to bear fruits in the fourth year when a small crop of 20–25 fruits may be harvested. The fruits mature in 130–180 days after fruit set, depending upon the season and location. The fruits are harvested when skin turns slightly yellow and fruits give a metallic sound when knuckled. When pressed, they give a crackling sound.

The fruit bearing increases with the age of the plant and 10-year old plants bear on an average 100–150 fruits weighing 20–30 kg. In well managed orchards, it is possible to get a good yield of 200–250 fruits/tree/year.

The sour fruits are harvested in October at about weekly intervals, cracked and seeds along with their flesh are taken out by hand and sun-dried on roof-tops or *pucca* platforms or on mats, until the desired moisture level is reached.

Yield

The average yield of different commercial varieties ranges from 10,000 to 12000 Kg/ha. However, 9 years old plants of Jyoti variety (G.K.V.K. 1) have recorded an yield of 250 fruits/tree/year (50.2 kg), with an estimated yield of 19,450 kg/ha.

Grading, Packaging and Storage

After harvest, the fresh fruits are graded according to the size and packed generally in bamboo baskets. In recent years, boxes made of light wood are used for packing premium quality fruits, and, individual fruits are wrapped in tissue paper.

Storage

Pomegranate fruits can be kept well for a long time. For bulk storage, the fruits are packed in layers in wooden crates containing 15–18 kg fruits and stored at 0 °C and 80% relative humidity (RH). Rice straw or paper-shreds are used as packing material.

The reddish-brown dried seeds (*Anardana*) are packed in gunny bags or in friction-free top tins and stored in a cool place.

Composition

The *Anardana* fresh fruit has the following composition (Table 3)

Preparation/Manufacturing of *Anardana*

Wild pomegranates have been processed by using different thermal treatments and three drying techniques with the following conclusions:

1. Deep sand roasting of fruits gave the best results for quick removal of seeds with flesh
2. The quality of solar-dried seeds was better than seeds dried by conventional drying methods
3. The drying ratio was 5:1 and rehydration ratio was 2.1:1 and for upgrading technology, the above processing techniques may be prescribed

Table 3 Composition of fresh *Anardana* seed

Property	Content
Moisture	78%
Protein	1.6%
Fat	0.1%
Fiber	5.1%
Other carbohydrates	14.5%
Mineral matter	0.7%
Calcium	10 mg/100 g
Magnesium	12 mg/100 g
Oxalic acid	14 mg/100 g
Phosphorus	70 mg/100 g
Iron	0.3 mg/100 g
Sodium	0.9 mg/100 g
Potassium	133 mg/100 g
Copper	0.2 mg/100 g
Sulfur	12 mg/100 g
Chlorine	2 mg/100 g
Vitamin A (Carotene)	0 mg/100 g
Vitamin B_1 (Thiamine)	0.006 mg/100 g
Vitamin B_2 (Riboflavin)	0.1 mg/100 g
Nicotinic acid	0.30 mg/100 g
Vitamin C (Ascorbic acid)	14 mg/100 g

Note: The above analysis is of fresh edible portion (seeds with flesh) of the pomegranate fruit constituting 68% of the fruit volume. Sample obtained from Coonoor in Tamil Nadu, India

Composition of *Anardana* (Dried Seeds with Pulp)

In the absence of any authentic data on the physico-chemical composition of *Anardana* in India, Pruthi (1987) carried out systemartic studies on the analysis of 40 samples of *Anardana* procured from various Divisional Forest Officers of Jammu and Kashmir State (19 samples), Himachal Pradesh (7 samples) and Punjab State (14 samples). The detailed results are presented in Tables 4 and 5 (Pruthi and Saxena 1984)

Quality Standards

There was considerable variation in color, moisture content, acidity and ash content among samples. In the absence of any national standard under the Agmark, ISI or PFAS at the commencement of this investigation, the authors (Pruthi and Saxena 1984) have proposed the following tentative standard for consideration of the Standard laying and Implementing Organization based "Frequency Distribution Histograms" on each item of comprehensive analysis.

Table 4 Proposed Indian Standards specifications for *Anardana*

Property	Limit	Percentage (m/m)
Foreign matter	Maximum	5.0
Moisture	Maximum	12.0
Acidity	Minimum	9.0
Total ash	Maximum	4.0
Acid insoluble ash	Maximum	1.2
Crude Fiber	Maximum	29.0

Source: Pruthi and Saxena (1984)

Table 5 Regional variations of the chemical composition of *Anardana*

	Jammu and Kashmir		Himachal Pradesh		Punjab	
Property	Range (19)	Average	Range (7)	Average	Range (14)	Average
Foreign						
Matter (%)	2.5–6.5	4.2	1.5–2.5	2.1	2.5–5.0	3.5
Moisture (%)	5.2–12.2	8.7	6.7–11.0	9.6	5.9–14.7	10.5
Acidity (%)	9.6–15.4	11.9	10.2–12.2	11.5	7.8–12.2	10.2
Total ash (%)	2.4–3.5	2.9	2.0–3.5	2.6	2.8–4.0	3.4
Acid insoluble						
Ash (%)	0.9–2.0	1.1	0.8–1.2	1.0	0.7–1.8	1.3
Crude Fiber (%)	22.6–30.0	26.6	22.0–26.0	24.2	26.5–29.0	27.6

Source: Pruthi and Saxena (1984)
Note: Figures in parenthesis are number of samples analyzed

Packaging and Storage

Based on systematic investigations conducted on the ERH, sorption isotherms, packaging in different types of containers and their storage at room temperature for 9 months, Pruthi and Saxena (1984) suggested that packaging and storage of *Anardana* in friction-top tins ensured good storage life. However, cost considerations favor HDPE (High Density Polyethylene) or double LDPE (Low Density Polyethylene) flexible pouches of 200–300 gauge.

Processed Products

The other processed products are pomegranate juice extracted from fleshy seeds, from which "*Anar Sharbat*" (soft drink) or "*Anar Syrup*" (fruit juice concentrate) are prepared and sold as soft drinks, which apart from their popularity, have also nutraceutical and pharmaceutical attributes.

Uses

Anardana is mostly used as a condiment for acidification of chutneys and certain curries, as s the case with *Amchur* and Tamarind.

The seeds are reported to be stomachic while the pulp is both cardiac and stomachic. The *Anardana* fruit rind decoction combined with cloves etc., is useful in treating diarrhoea and dysentery. The rind is also used as dyeing material for cloth. Seeds are also reported to have estrogenic activity as they contain steroidal estrogens.

Stem bark and root bark are astringent, anthelmintic, and specific in tapeworm treatment. They contain beneficial alkaloids.

References

Pruthi JS (1987) Anardana. In: Spices and condiments, 3rd edn. National Book Trust, New Delhi, pp 14–16

Pruthi JS, Saxena AK (1984) Studies on *anardana* (Dried pomegranate seeds): Composition and packaging requirements of *anardana*. J Food Sci Technol 21(5):296–299

Rao M, Purushotham, K (1988) Fruit borer menace in pomegranate. Spices News L 22(8):17–18

Balm or Lemon Balm

Abstract The chapter covers, at first, description, distribution, area, production and exports. Additionally, it would also discuss crop management, post harvest technology and processing technology. Special emphasis will also be placed on the use of end products.

Keywords Description · Distribution · Crop management · Post harvest technology · Processing technology

Botanical Name: *Melissa officinalis* Linn.
Family: Labiatae

Description and Distribution

Balm or Lemon balm is a perennial herbaceous plant of the mint family, with a strong, agreeable odor, reminiscent of lemon and that is why it is called lemon balm. It is evergreen, 30–60 cm in height, and grows for 2 years or more. In some cases even longer. According to Guenther (1972), it can last upto 10 years, but, is normally, maintained only for 4 years. The blossoms are small and of white or light rose color. The leaves or the herb is used as a spice and as a flavorant. A native of the countries bordering northern Mediterranean, it grows wild and is also cultivated in gardens as a medicinal herb. It has been naturalized in the USA, grows wild in the eastern USA, and is found in the temperate Himalayas. Another plant *Melissa parviflora* Benth., which is considered a good substitute for *M.officinalis* (lemon balm), is an erect, pubescent or glaborate herb, 60–100 cm high, found in temperate Himalayas from Garhwal to Sikkim, Darjeeling, Khasi, Aka and Mishmi Hills, at an altitude of 1200–3000 mm. Leaves are ovate or ovate-lanceolate. Flowers are white or pale-pink, rarely yellow, in a few or many flowered auxiliary whorls; nuts are narrowly obovoid, dark, rugulose. Balm is now well established in Himachal Pradesh (Solan) (Meenu Sood et al. 1996).

K. P. Nair, *Herbal and Acidulant Tree Spices*,
https://doi.org/10.1007/978-3-031-04360-4_4

Crop Management

Soil and Climate The Melissa plant is easily affected by cold, hence, a sunny and warm location should be chosen to grow the plant. Although it will grow almost anywhere, deep and shaded soils of medium consistency (which can be irrigated) are preferable. The plant will thrive well in a fresh and fertile alluvial land, but, will suffer from excessive moisture. In light and dry soils, the leaves turn yellow and the yield gets greatly reduced.

The soil should be prepared to a depth of about 30–35 cm. during second tilling, organic manure and chemical fertilizers should be applied.

Propagation The plant can be propagated either from seed or from cutting. In places near Paris, the first method is usually preferred. The seed is planted in July in greenhouse, and the young plants are placed into their permanent position in October. In other regions, the seeds are planted directly into the permanent place in April. The planting of the cuttings takes place in the fall or in March-April, 50–60 cm apart.

Cultural Practices They comprise irrigation, repeated turning over of the soil and seeding. Although the plant requires moisture, excessive watering in the soil is harmful. At the beginning of the winter, land should be tilled, fertilized, and in a cold climate the soil should be mounted at the base of he plants. The pants may last for about 10 years, but, are maintained for only 5 years for the purpose of obtaining the largest yield. After that period, the crop is rotated with others. Application of 90 kg N/ha gave mean maximum leaf yield of 127.6 kg/ha and essential oil yield of 28.4 l/ha. A row spacing of 30 cm × 45 cm was found optimum for higher yield. Wider spacing resulted in lower yield and growth (Meenu Sood et al. 1996).

Harvest When plants are in full bloom, the crop is cut with a scythe or mower. In the first year, the harvest takes place in August, when the flowers are well formed; thereafter, two cuttings can be made, one in June, another in August. It is advisable to harvest on a clear day, after the dew has evaporated, or else, moist leaves would turn black on drying. The plants are very susceptible to heat and sunlight, and hence, should be transported as soon as possible to the distillery or artificial drier if the leaves are to be used for liquor distillation. In this case, the leaves are cleaned and dried, cleaning being accomplished after stalks of the suspended plants dry.

Yield In Anjou (France), about 1800 kg/ha of cleaned and red leaves are obtained per hectare. They are sold in the market at attractive prices to liquor industry, to the pharmaceutical trade and to herborists.

Post Harvest Technology If flowering tops are to be marketed, the stemmy portions must be removed before or after drying. Drying is done in shade in order to preserve its natural color. For manufacturing essential oil, no drying is necessary.

Fresh flowering tops and leaves immediately after harvest give higher yield of essential oil.

Processing Technology

Manufacturing of Essential Oil The volatile oil obtained from fresh herb is very low (0.01–0.014%). That is why this oil is expensive. The commercial oil, therefore, is seldom, if ever, pure. It consists of mixtures; largely of lemongrass and citronella oil, of fractions thereof. In European countries, it is sometimes prepared by distilling lemon oil over *Melissa officinalis* Linn, herb (*Oleum melissa citratum*). By distilling fresh *Melissa officinalis* at the beginning of the flowering stage, Sehimmel and Co. could obtain 0.014% of true oil, with a specific gravity (15°) of 0.924 and an optical rotation of + 030.

The fresh herb during full-bloom gave 0.010% of oil, with specific gravity (15°) of 0.894; the oil was optically inactive. The odor suggested the presence of citral and citronella; the scent of the first mentioned oil was more agreeable, typical of *Melissa*.

The following table (Table 1) details the physico-chemical properties of Balm oil.

Uses

As a Food Flavorant Fresh or powdered balm leaves are used in fish dish stuffings or as a substitute for lemon. Fresh leaves are piquant in salads and in summer drinks. Thus, balm leaves are widely used in culinary flavoring.

In Medicine Leaves and flowering tops are also used in medicine. Lemon balm is said to possess stomachic, anti-tubercular and anti-pyretic properties. It is used to

Table 1 Physico chemical properties

Property	Content
Specific gravity at 25 °C	0.8910
Optical rotation at 22 °C	$+2^0$ 8 °C
Refractive Index at 22 °C	1.4704
Acid number	2.2
Ester number	27.42
Ester number after acetylation	236.28
Aldehyde content (bisulfite method)	42.0

Solubility in 0.5 vol of 90% alcohol and in 2 vol of 80% alcohol strengthen gums and to remove foul taste from the mouth. The fruit is considered a brain tonic and is useful in hypochondriac condition. Leaves and stems are also said to be useful in brain, liver and heart diseases and also bites of venomous insects (Pruthi 1998).

The volatile oil distilled from the plant is used to flavor and also somewhat in perfumery. However, this oil is not popular on account of its high price (Guenther 1972). It may be possible to simulate its odor by blending it with other oils like lemongrass oil or citronella oil (Baun 1942).

The balm oil is reported to be sedative. Its properties resemble those of peppermint oil and balm distillates can act as mild spasmodic agents.

References

Baun (1942) Fortsch. Therap 18:42

Guenther E (1972) The essential oils, vol III. D. Van Nostrand Co, New York, pp 395–399

Meenu Sood, Singh JM, Srivastava LJ, Rana RC (1996) Studies on the response of *Melissa officinalis*. Linn. (Gulanga) to nitrogen and row spacing. Indian Perf 40(4):107–109

Pruthi JS (1998) Spices and condiments, 5th edn. National Book Trust, New Delhi, pp 1–300

Basil or Sweet Basil (*Tulsi*)

Abstract The chapter starts with the description and distribution of the plant and covers other aspects, such as, minor types of basil, such as French basil, Indian basil etc. It would also cover area, production and exports. Special emphasis would be placed on end products and quality of oil.

Keywords Description · Distribution · Oil quality and end uses · French basil · Indian basil seeds

Botanical name: *Ocimum basilicum* Linn.
Family: Labiatae

Description and Distribution Basil, French basil or Sweet basil, Tulsi or *niyazbo* (*Ocimum basilcum* Linn.) is an erect, glabrous herb, 30–90 cm high, native of North-Western India ad Iran, is an annual of the mint family and constitutes an important culinary herb. It is a rich source of valuable essential oil. In India, there is a widespread belief that if planted around homes and temples, it ensures happiness. Hindus consider it sacred and also good for health when its fresh leaves are eaten raw and also drink their decoction. It is indigenous to lower hills of the Punjab and Himachal Pradesh and is cultivated throughout India. It is now cultivated in Southern France, Egypt, Belgium, Hungary and other Mediterranean countries, and also in the USA. The plant grows to a height of about 60 cm. The freshly picked bright green leaves measure up to 3.75 cm in length. When dried, they turn brownish green, whole and broken brittle, curled or folded together. Dried leaves and tender 4-sided stems of this plant are used as a spice to flavor and to recover essential oil in them. The flavor is warm, sweet and somewhat pungent and peculiar. The odor of sweet basil is aromatic, fragrant and sweet.

The leaves have numerous dot-like oil glands in which aromatic volatile oil of the herb is contained. The herb bears clusters of small, white-lipped flowers in raceme fashion. It can easily be grown at home or in gardens in ordinary soil.

K. P. Nair, *Herbal and Acidulant Tree Spices*,
https://doi.org/10.1007/978-3-031-04360-4_5

The plant is very variable and its botanical nomenclature is complicated; several designations have often been assigned to one or the same type. Polymorphism and cross-pollination under cultivation have given rise to a number of subspecies, varieties and races, differing in height, habit of growth, degree of heaviness, and color of stems, leaves and flowers, and types have been confused with other species.

"The smell of basil is good for the heart and head" wrote the seventeenth century herbalists, "and it maketh man merry and glad". This delightful annual herb of the mint family is native to India and Persia (Iran). In France, it is called the "*herbe royale*", and its aroma is highly esteemed in that food-loving nation.

Major Types of Basil

American Basil Commercial sale production of the Basil in the USA began shortly after the World War II. Basil is a plant which lends itself well to modern American farming and processing techniques, mostly in California. Consequently, the domestic Basil is highly prized for its rich color, sweet flavor, cleanliness and uniformity of milling, to the point where it typically brings as much as thrice more the price of the most expensive imported Basil. Uniformity is, in particular, important to spice packers because it assures them of even attractive fills in their containers. The domestic product is mechanically dried and then milled into 4 particle sizes: coarse, primarily for retail packs, medium, popular in food service and salad dressing manufacture; fine, much used in the pizza industry; and ground, or powdered, for use in any product where aroma and flavor are important without any evidence of leaves. The choice of leaf particle size is unique with the American product. Coarse cut runs about 0.63 cm, medium about 0.27 cm, fine about 0.15 cm in size.

French Basil The French Basil is the most expensive of the imported products. It is characterized by the trade as being significantly better (somewhat sweeter) in flavor, deeper green in color, and cleaner in appearance than any other imports. It ranges from 0.5% to 1.1% in volatile oil, with the average of about 0.75%. Most of the commercial Basil in France is a product in the Province region.

Egyptian Basil In recent years, Egypt has been the largest exporter of Basil to the USA. Egypt is no newcomer to Basil production. But for a long time, most of the output has been shipped to Europe. The Egyptian product is said, by buyers, to have a "mintier" flavor than other sweet Basils. It is priced considerably more under the name French Basil in Egypt, and is produced mainly in the regions of the Beni Suif and Fayon in the vicinity of the Nile river.

Indian Basil It has already been described earlier. It is also called French Basil or Sweet Basil. In 1964, the USA was importing only 40,000 pounds of Basil per annum. Today the US annual import exceeds 1 million pounds (25 times more) and it is joined by domestic production up to 75%. These figures become all the more impressive when we consider that a pound of Basil will season enough Chicken Cacciatore to feed over 3000 persons. The Basil boom is another tribute to the popularity of the pizza and Italian foods, in general. Many pizza makers today are using Basil in their tomato sauce and then sprinkling Oregano over the top of the finished pie. Pizza brought Oregano out of the realm of the unknown to an annual consumption of over 5 million pounds! And now, it seems well on the way to doing the same to Basil.

The flavors of Basil and tomatoes have a special affinity and this is the basis of today's increased usage of this herb in the Western World. America has adopted Southern Italy's love for tomato sauce. At first, their attention was turned to Oregano because of its high profile as a topping on pizza. More recently, they have discovered that good Italian cooks often use Basil before Oregano – that is, in preparing the basic sauce for the pizza (and for that matter, just about any sauce). Basil thus becomes "Mr. Inside" and Oregano "Mr. Outside", as a football fan would describe it. This has led to a wide range of Basil in all sorts of tomato-sauce recipes, from spaghetti sauces to soups, tomato juice cocktails, tomato-based salad dressing and almost all the countless number of ethnic specialities.

Area, Production and Exports

No statistics are yet available in India on these aspects. However, there is considerable commercial-scale production of Basil in USA, France, Hungary, Egypt, Bulgaria, West Germany, Poland, Yugoslavia, Belgium, Turkey, Italy and The Netherlands. The USA imported 412 tonnes of Basil worth US $ 400,000. No production data are available for any of the countries named above. In India, however, Basil oil is now being produced on a commercial scale to a limited extent.

According to Varshney (1992), about 200 tonnes of Basil oil was produced in India as compared to 250 tonnes in the world trade. Earlier, India produced only about 2 tonnes of oil. This is a remarkable jump in Basil cultivation and oil production from the plant.

Crop Management There are numerous varieties of *Ocimum basilicum*, of which 4 are identified in India. They are as follows:

1. *Ocimum basilicum* Linn. var. *album* Benth. (lettuce-leaf basil)
2. *Ocimum basilicum* Linn.var. *differme* Benth. (curly-leafed basil)
3. *Ocimum basilicum* Linn. var. *purfurascans* Benth. (violet-red basil)
4. *Ocimum basilicum* Linn. var. *thyrsiflorum* Benth. (common white-basil)

Curly-leafed basil is considered most suitable for cultivation. It is grown in France and is reported to give good yields of high quality oil.

Soil and Climate Basil can be grown in a wide range of climates and soil. It is very versatile in its adaptation to climate and soil types.

Breeding Tesi et al. (1991a) have investigated breeding of basil ad varietal evaluation and also its floral biology and reproduction (Tesi et al. 1991b). Gupta (1994) analyzed the genetics of some chemo-types of basil var. *glabratgum*.

Propagation Sweet basil *Ocimum basilicum* is propagated by seeds and is commonly grown in gardens as an aromatic herb. The best season for sowing it in the plains of India is October–November, and, in the hills it is March–April. Seedlings are raised in the nursery beds and transplanted 30 cm apart in rows spaced at 40 cm apart. The crop is ready for harvest in 2–3 months time after planting. Davis (1992) has made comparisons of mulches for fresh market basil production.

Plant Protection

Diseases The first reports of *Fusarium* wilt was made by Davis et al. (1993). Keinath (1994) investigated pathogenicity and host range of Basil for Fusarium and the evaluation of the control methods thereof. El Massy et al. (1995) have reported the effects of herbicides on sweet Basil and Marjoram.

Harvest Several (3–5) cuttings of leaves and flowering tops may be made during the season. Plants are cut close to the ground, bunched and dried. The dried leaves and flowering tops are stripped from stems and packed in closed containers. In Assam, workers have shown that harvesting of plants should be done after the 12th week from sowing the seeds, hereafter at intervals of 2 months each. In all, 3 harvests are possible.

In the *Terai* region of Uttarakhand in India the optimum yield of herb is obtained by taking only the first 4 crops of flowers (main sub-inflorescence) and the last crop of the entire flowering herb. The first harvest is taken when plants are in full bloom, second and subsequent harvests are available thereafter after at every 15–20 days interval. The last harvest is taken of the entire plant and distilled. While harvesting the crop, precaution should be taken that the root system of the plant is not injured, otherwise yield of subsequent harvests will be adversely affected. The crop is to be harvested at the stages of maximum bloom to seed set (Gupta 1996).

Yield Yield up to 6800 kg/ha of leaves and flowers in 2 cuttings/harvest is reported from trial cultivation in Kanpur, North India. In the USA, yield up to 20–25 tonnes/ha of fresh herb/ha has been reported, and, 3–5 harvests per annum can be had. Kimura et al. (1994) have reported the effects of watering/irrigation on growth, yield, essential oil concentration and evapotranspiration of Basil.

Post Harvest Technology

The post-harvest shelf-life of sweet Basil has been investigated in good detail by Lage ad Cameron (1994). Rocha et al. (1993) have reported the effect of pre-treatments and drying conditions on the drying rate and color retention in Basil.

Packaging and Storage The dried leaves are generally packed in tin containers kept closed. The enzymatic browning of Basil has been attributed to oxidation of phenolc compounds by polyphenol oxidase, naturally present therein (Baritaus et al. 1991).

The following table (Table 1) gives details of the composition of sweet Basil.

A good commercial sample of sweet Basil has been found to contain the following:

Volatile oil (minimum) 0.4%
Total Ash (maximum) 15%
Acid insoluble ash (maximum) 1%
Moisture (maximum) 8%
Total Ether extract (minimum) on moisture-free basis 4%

Considerable variations in the various physico-chemical properties of Basil oil produced in different Indian regions, such as, Jammu, Uttar Pradesh (now Uttara Khand), Assam are shown in the following table (Table 2).

Table 1 Chemical composition of Basil as per the specifications of ASTA, USA

Details	Content
Moisture	6.1%
Protein	11.9%
Fat (Ether extract)	3.6%
Fiber	20.5%
Carbohydrates	41.2%
Total ash	16.7%
Calcium	2.1%
Phosphorus	0.47%
Sodium	0.04%
Potassium	3.7%
Iron	0.04%
Vitamins (mg/100 g)	
Vit B_1 (Thiamine)	0.15
Niacin	6.90
Vit B_2 (Riboflavin)	0.32
Vit C (Ascorbic acid)	61.3
Vit A	290 IU (International Unit/100 g)
Calorific value (Food energy)	325 calories /100 g of dried herb

Note: American Spice Trade Association (ASTA), USA

Table 2 Variations in composition of Basil oil

Details	Jammu oil	UP oil	Assam oil
Oil yield	0.5–0.6%	0.1–0.28%	NA
Specific gravity	0.9016 (20 °C)	0.886–0.9288	0.9183 (29 °C)
Refractive Index	1.4892 (20 °C)	1.4662–1.4757	1.4798 (29 °C)
Optical rotation	-10^0 7	$-4-7^0$	-9.0^0
Ester value	1.2	4.60–7.10	7.20
Alcohol (%)	48.5	41.10–50.0	54.70
Methyl chavicol (%)	–	35.20–40.50	54.70

Source: Sharma et al. (1987); *NA* Not Available

Processed Products

The main product manufactured from Basil leaves and flower tops is the essential oil known as "Oil of Basil", which is discussed below. Two types of Basil oils are mainly recognized in the world market – The European and the Reunion (Guenther 1972).

Distillation of Oil

Oil sweet Basil is produced by the hydro-distillation of the herb. The flowers or whole herbs or both are packed into the distillation unit and hydro-distilled or steam-distilled. It takes, in all, about 4 h to complete 1 charge. The oil, being lighter than water, is easily separated from the oil-water mixture. It is advisable to use distillate after removing oil for further charge, as it contains small quantity of oil. Oil yield thus increases. Precaution should be taken that the distillation unit is clean and free from other odors, or else the oil will be contaminated with undesirable odors and colors. It is possible to produce oil worth about INR 10, 000 (1 US $ = 75 INR) from one hectare during June–November. The expenditure involved is about INR 2500 to INR 3000/ha. Two grades of oil are produced, depending on the part of the plant harvested.

Yield of Herb Oil

The flowers, on an average, yield 0.4% oil, while the whole plant (Indian Basil) contains 0.10–0.25% oil. By taking three harvests of the herb at 2-months interval about 15 tonnes of herb/ha can be had, corresponding to 30–35 kg oil. By taking the initial 3 or 4 harvests of flowers (including the main and sub-inflorescence) and final harvests of the whole herb, about 3–4 tonnes of flowers and 113 tonnes of whole herb can be had/ha corresponding to about 13 kg of the flower oil and about 27 kg of whole herb oil, respectively; in all, 40 kg of oil/ha is obtained. In case of labor shortage for harvest, 3 harvests of whole herb at an interval of 1 month can be had after the first harvest which gives about 15 tonnes of herb corresponding to about 30 kg of oil. Oil of sweet basil, produced from both herb and flowers is saleable. In the long run, oil from the whole herb is economical to produce.

Quality of Oil

The quality of the essential oil is correlated to the composition of the oil, which is greatly influenced by the climatic as well as agronomic factors. Out of the various important agronomic factors which influence herb and oil yields and oil quality, the time of transplanting and stage of harvest are the most important. Randhawa et al. (1994) have investigated these aspects, and the investigations revealed that to get maximum herb and oil yield, French Basil should be transplanted from end March to first week of May. The oil content in the herb was maximum at the vegetative stage. The refractive index and the optical rotation of the herb oil increased with the delay in planting from March 10 to May 10 and March 10 to March 25 respectively, and beyond March 25 optical rotation decreased progressively. Each delay in harvest from vegetative to complete flowering stage, increased optical rotation and refractive index of the herb oil, but, the ester value decreased significantly with delay in harvest.

The oil obtained from flowers is better than the oil from the whole herb in quality. Both oils are saleable (from flowers and whole herb), and the yield and quality are comparable to those produced overseas. A sample of oil distilled from plants grown in Chalakkudi (Kerala State, India) had a fine lavender odor and was rich in linalool and methyl cinnamate. Another sample obtained from plants grown at the HB Technology Institute, Kanpur, Uttarakhand State, India, contained methyl cinnamate, linalool, methyl chavicol and ocimene. The quality of volatile oil varies greatly in composition and properties owing to differences in species, soil, climate and the part of the plant used. The European oil, known to be true to the name, shows according to results of several authors, specific gravity at 15 °C: 0.895–0.930; refractive index: 1.477–1.495; optical rotation: −22° to 6°; ester number 3–15; linalool as 34.50–39.66%. The methyl chavicol content of all these oils was about 55%, calculated by determining the methoxy number according to Zaisel's method.

Methyl chavicol, the main constituent of sweet Basil oil, is oxidized on ageing and on exposure of oil to light and air; thus, older oils usually show a higher specific gravity and higher refractive index. Sweet Basil oil must, therefore, be stored carefully.

Adulteration of Sweet Basil Oil The oil is frequently adulterated with the much lower-priced Reunion Basil oil. This is indicated by increased specific gravity and refractive index and by a lowered levo-rotation or slight dextrorotation depending upon the amount of Reunion oil present. These discrepancies can be partly corrected by the addition of 1-linalool, therefore, adulteration of sweet Basil oil is not always detected by mere routine analysis. The expert will submit the oil to a careful organoleptic test, watching especially for the presence of camphor, which does not incur in true sweet Basil oil.

Uses

As a Food Flavorant Sweet Basil is used to flavor numerous foods – soups, meat pies, fish, certain cheeses, tomato cocktail, zucchini, eggplant, cooked cucumber dishes, cooked peas, squash and string beans: chopped Basil is sprinkled over lamb chops before cooking. Basil is often used with, or, as a substitute for oregano in pizza topping, spaghetti sauce or macaronic and cheese casseroles. It is also used in the manufacture of Chartreuse and other liqueurs. Sweet Basil from Italy and America is known for its quality. In Italy, Basil is an important seasoning in tomato-paste products. In France, it is known as the "herbe royale".

Although not used in large quantities, the oil of sweet Basil is employed quite extensively in all kinds of flavors, including those for confectionery, baked goods and condimentary products (chilli sauce, catsups, tomato pastes pickles, fancy vinegars) and in spiced meats, sausages, etc. The oil also serves to impart distinction to flavors in certain dental ad oral products.

In Perfumery and Cosmetics It is used in certain perfumery products. For scenting soaps, the lower priced Reunion oil is preferred.

In Medicine The plant is a stomachic, anthelmintic, alexipharmic, antipyretic, diaphoretic, expectorant, carminative, stimulant and pectoral. An infusion of the plant is given for cephalalgia and gouty joints, and used as a gargle for foul breath. The plant is used in homeopathic medicine, as well.

Roots, bark and leaves are cyanogenetic. Alcoholic extracts of leaves and flowers possess antibacterial activity against *Micrococcus pyrogenes* var. *aureus*. Seeds possess demulcent, stimulant, diuretic, diaphoretic, and cooling properties.

Insecticidal and Bactericidal Properties Basil oil possesses both insect repellant and bactericidal properties.

Uses of Basil Seeds Seeds are colorless with an oily, slightly pungent taste. When steeped in water, they liberate a mucilage which is semi-transparent and nearly tasteless. The mucilage (9.3%) yields on hydrolysis uronic acid, glucose, xylose and rhamnose. The seeds contain a drying oil with the following fatty acids composition: palmitic: 7%, stearic: 0.2%; oleic: 11%; linoeic: 60%; linolenic: 21%. The unsaponifiable fraction is reported to contain beta (please use symbol) sitosterl, olenolic and ursolic acids.

Aqueous extract of seeds are active against gram-positive bacteria and *Micrococus pyrogenes* var. *aureus*. Infusion of seeds is used to treat gonorrhea, dysentery and chronic diorrhea.

References

Baritaux O, Amiot MJ, Richard H, Nicholos J (1991) Enzymatic browning of basil (*Ocimum basilicum* Linn.) studies on phenolic compounds and polyphenol oxidase. Sci Aliment 11(1):49–63

Davis JM (1992) Comparison of mulches for fresh market basil production. Hort Sci 29(4): 267–269

Davis RM, Marshal KD, Valencia J (1993) First report of *Fusarium* wilt of basil in California. Plant Dis 77(5):537

El Massy MH, Charles DJ, Simon JE (1995) Bentazon and terbaeil as post-emergent herbicides for sweet basil and sweet marjoram. Int J Geogr Inf Syst 3(3):19–26

Guenther E (1972) The essential oils, 2nd Reprint D. Van Nostrand Co., New York

Gupta SC (1994) Genetic analysis of some chemotypes in *Ocimum basilicum* var. *glabratum*. Plant Breed 112(2):135–140

Gupta SC (1996) Phenological observations on basil. Indian Perfumer 40(1):17–22

Keinath AP (1994) Pathogenicity and host range of *Fusarium oxysporum* from sweet basil and evaluation of disease control methods. Plant Dis 78(12):1211–1216

Kimura M, Ichimura M, Tomitakay (1994) Effects of watering on growth, yield, essential oil concentration and evapotranspration of sweet basi. Japanese J Trop Agri 38(1):65–72

Lange DD, Cameron AC (1994) Post-harvest shelf-life of sweet basil. Hortic Sci 29(2):102–104

Randhawa GS, Gill BS, Saini SS (1994) Effect of different transplanting dates and crop growth stages and physico-chemical properties of French basil (*Ocimum basilicum* L.) oil. Indian Perfumer 38(4):123–128

Rocha T, Lebert A, Martyaudouin C (1993) Effect of pretreatments and drying condition on drying rate and colour retention of basil. Food Sci Technol 26(5):456–464

Sharma A, Tewari R, Virmani OP (1987) French basil – a review. Cromap 9(3):135–151

Tesi R, Paolucci B, Nencini A (1991a) Breeding basil (*Ocimum basilicum* L.) I. Floral biology and reproduction. Sementi Elette 37(3–4):31–35

Tesi R, Paolucci B, Joshi D, Vidriech V (1991b) Breeding basil (*Ocimum basilicum* L.) II. Varietal evaluation. Sementi Elette 37(5):7–13

Bay or Laurel Leaves

Abstract The chapter will at first discuss the description and distribution and move on to other aspects of production, such as post harvest technology, processing technology, essential oil recovery, and end product uses.

Keywords Description · Distribution · Post harvest technology · Processing technology · Essential oil

Botanical Name: *Laurus nobilis* Linn
Family: Lauraceae

Description and Distribution

Laurel leaves or Bay leaves are dried leaves of *Laurus nobilis* Linn. an evergreen tree or shrub. The plant is grown in the Mediterranean countries and is cultivated in Greece, Spain, Portugal, Asia Minor and central America. It is sometimes grown in Indian gardens but does not seem to thrive well there. The surface color of the leaf is green; the underside is pale-green and somewhat yellowish. Laurel leaves are used whole or cracked. The aroma of the crushed leaves is delicate and fragrant and tase is aromatic and bitter.

The leaves should not be confused with the leaves of bay rum tree [*Pinneta racemosa* (Mill) J.W.Morre], belonging to the Family Myrtaceae, from Puerto Rico and nearby islands, or with the California bay laurel (*Umbellularia californica* Nutt. family Laureaceae). The sweet bay or laurel leaves originate from an evergreen hardy tree or bush; cultivated since antiquity in Mediterranean countries. Size of the leaves is variable, ranging from 2.5 to 7.5 cm or more in length and 1.6–2.5 cm or more in breadth, at the widest part of the leaf. The shape of the leaf is elliptical, tapering to a point at the base and tip of the leaf.

K. P. Nair, *Herbal and Acidulant Tree Spices*,
https://doi.org/10.1007/978-3-031-04360-4_6

Dried berries, commonly called bay berries, have been imported into India for medicinal use. The berry is ovoid (about 1.5 cm long), black, coarsely wrinkled and contains a single seed.

Post harvest Technology

Drying Laurel trees are pruned from September to March and the leaves are dried as usual in shade to the desired moisture level. The leaves are esteemed for their excellent flavoring property.

Composition According to ASTA, the chemical composition of the dried bay leaves is as follows:

Moisture: 4.5%
Protein: 7.6%
Fat: 8.8%
Fiber: 25.2%
Carbohydrates: 50.2%
Total ash: 3.7%
Calcium: 1%
Phosphorus: 0.11%
Sodium: 0.02%
Potassium: 0.6%
Iron: 0.53%
Vitamins mg/100 g:
Vit B1 (1 subscript) thiamine: 0.10%
Vit B2 (2 subscript) riboflavin: 0.42
Niacin: 2.0
Vit C (Ascorbic acid): 46.6
Vit A: 545 IU (International Units)
Calorific value (food energy): 410 calories/100 g
Leaf analysis (made at Munster Experiment Station, Berlin):
Moisture: 9.45%
Protein: 8.34%
Fixed oil: 4.49%
Volatile oil: 3.63%
Alcohol extract: 25.01%
Nitrogen-free extract: 38.33%
Fiber: 31.83%
Ash: 4.53% (Italian variety)
Pentosans: 13.84%

Processing Technology

Essential Oil Recovery The leaves on steam distillation yield 1–3% essential oil with a characteristic sweet and spicy odor, reminiscent of cajuput. According to another report, fresh leaves and terminal branchlets yield 0.5% oil, while dried leaves yield about 0.8%. Guenther (1972) has reviewed the published variation in composition from different countries.

Physico-chemical Properties of Volatile Oil

According to Parry (1969) Laurel leaves yield 1–3% of volatile oil, having the following properties:

Specific gravity at 15 °C: 0.915–0.930 (occasionally higher)
Optical rotation at 20 °C: 15 –22 °C
Refractive index At 20 °C: 1.4670–1.4775
Principal constituent: Cineol 25–50%
Solubility: 1 part in 3 parts of 80% alcohol.

Its principal constituent, up to 50%, is cineol, a colorless liquid with a strong aromatic, camphoraceous odor, and a cooling taste. Other organic compounds include α-pinene, α-phellandrene, 1-linalool, 1-β-terpineol, geraniol, eugenol, eugenol acetate, methyl eugenol, a number of esters, acetic, isobutryric and isovaleric acids.

Uses

Bay Leaves for Flavoring Bay leaves are among the oldest herbs of the word. Wreaths of laurel leaves were used by Greeks and Romans to honor their heroes. Laurel leaves are used principally in vinegar pickle when packaging pig's feet, lamb and pork tongue. They are also used to flavor soups, stews, meat, and game dishes, fish and sauces, pickling spice, and also in confectionery. Laurel leaves are available whole or cracked, are not usually ground, and are packed in 50 lb and 110 lb capacity cartons.

Both leaves and fruits, possessing aromatic, stimulant and narcotic properties, were, formerly employed to cure hysteria, amenorrhoea and flatulent colic. They are even used internally, though rarely, at present. Externally, however, commercial oil of laurel berry is sometimes applied as a stimulant in sprains, but, its principal use is in veterinary medicine.

Fat from Bay or Laurel Leaves

Extraction of Berry Fat The berry (pericarp 30%, seed 70%) yields 30–34% of an aromatic fat used to some extent in pharmacy, veterinary practice and perfumery. Commercial fat is obtained from whole berry by pressing or boiling in water and skimming off the separated fat.

Use of Berries in Medicine Berries are used in medicine to treat diarrhoea and dropsy. In Europe, they are also used to promote miscarriage.

Laurel Wood Utilization Laurel wood resembles walnut wood in grain and color and is suitable for decorative cabinet work.

References

Guenther E (1972) The essential oils, vol I–VI (Reprinted). D. Van Nostrand Co, New York

Parry JW (1969) Spices II. Their morphology, histology and chemistry. Chemical Publishing Company Inc. New York

Kokam or Kokum

Abstract The chapter will at first discuss description and distribution and then move on to other aspects, such as, area, production and exports, crop management, flowering and fruiting and conclude with details of post harvest technology and processed products.

Keywords Description · Distribution · Crop management · Post harvest technology · Processed products

Botanical name: Garcinia indica Choisy
Family: Guttiferae

Description and Distribution

Kokam or Kokum also known as "Kokam Butter Tree", "Brindonia Tallow Tree" or "Mangosteen Oil Tree" should not be confused with "Mangosteen" (*Garcinia mangostana*), another species of the same genus *Garcinia*. Kokam is a slender evergreen tree with drooping branches; leaves are ovate or oblong lanceolate, 6.25–8.75 cm long and 2.5–3.75 broad, dark green above and pale beneath; fruits are globose or spherical, 2.5–3.75 cm in diameter, dark purple when ripe, enclosing 5–8 large seeds. The tree is found in Southern India, in particular, in the state of Karnataka, India and in the tropical rain forests of the Western Ghats, from Konkan, (Maharashtra State) Coorg (Karnataka State) and Waynad (Kerala State). It is often planted in the southern districts of Maharashtra State, notably, in Ratnagiri district and to some extent in Gujarat State. It is reported that it flourishes well on the lower slopes of the Nilgiris hills in Tamil Nadu. It also grows in West Bengal State and Assam State. It flowers from November to February and fruits ripen in April–May.

The Kokam of commerce comprises the sun-dried rind of the fruit, repeatedly soaked in its juice. It is traditionally (Indian tradition) used as an acidulant in some foods and curries.

K. P. Nair, *Herbal and Acidulant Tree Spices*,
https://doi.org/10.1007/978-3-031-04360-4_7

Area, Production and Exports

No reliable recent published statistics is available on the area, production and export of *Kokam*. However, according to the DMI report (1968), 920 tonnes and 15 tonnes of *Kokam* were produced in Maharashtra and Gujarat states respectively. In the Konkan region alone, of Maharashtra State, about 5000 tonnes were produced. According to a survey report, Western Ghat region contains about 1,500,000 *Kokam* trees, with an estimated yield of 10,000 bags each, containing 93.31 kg of seeds (Muralidharan 1976). It is mostly consumed within India.

Kokam Exports

During 1987–88, about 74.91 tonnes of *Kokam* were exported from India at a total export value of INR 2.22 million, at an average FOB (Freight On Board) unit value of INR 29.72/kg of *Kokam*. However, during 1990–91, about 55,770 kg of Kokam worth INR 3 million was exported from India (following table, Table 1). Export figures are given in Table 1 and export trends are given in Table 2.

Table 1 Export of *Kokam* from India

Export destination	Quantity (kg)	Value (INR)
Canada	1374	53,443
Hong Kong	160	7864
Japan	45,000	2,686,783
Oman	286	8673
Pakistan	2000	5920
United Arab Emirates	600	12,600
U.K.	2650	78,150
U.S.A.	3706	153,191
Total	65,776	3,006,624

Table 2 Export trends of Kokam from India

Year	Quantity (tonnes)	Value (INR, 000)
1988–89	40.7	954.2
1989–90	271.4	7705.2
1990–91	66.9	3271.7
1991–92	14.3	481.0
1992–93	12.2	496.2

Source: Spices Board, Kochi 1994. Export Statistics 2nd edn. pp. 1–162

Crop Management

Standard book of horticulture by Naik (1949) covers this aspect of *Kokam,* which may be referred to, when required. However, a brief account of its flowering, fruit-formation, harvesting and related problems are discussed here.

During 1991–92 and 1992–93, 14.3 and 12.2 million tonnes of *Kokam* was produced and exported worth INR. 4.81 lakh (1 lakh = INR 100,000) and INR 4.96 lakh respectively. This export trend is being kept up during 1993–94 onwards, since *Kokam* is becoming quite popular, as an acidulant spice used in curries.

Flowering and Fruiting

Dioecious by nature, the productive *Kokam* female trees outnumber male ones (referred to as "*varangu*" in vernacular), which are usually not retained after they have exhibited their sexuality.

Every year, from February to April, the trees put forth inconspicuous, pale-yellow flowers which after fruit-set develop into ovoid, ellipsoid or spherical fruits. The fruits vary in size, weighing 50–180 grams, and, the fruit rind is grooved into 7–10 segments. On pricking, they exude a yellow resinous liquid which hardens slowly into a brown gummy mass.

Harvesting and Post-harvest Treatments

Fruits abscise on ripening and are collected, and the seeds, enclosed in a pulpy transluscent aril, which is often sweet and edible, but, sometimes acrid, are separated from the rinds. The rinds are then dried. The seeds take more than 7 months to germinate.

Owing to the coincidence of the south-western monsoon rains with the period of fruit ripening, farmers resort to smoke drying of rinds in lofts above fire places. In a week or two, the dry rinds attain a coal black color and a characteristic acrid taste. The blend of this with the acid components imparts an exceptional flavor to the dried rinds.

The rinds are then preserved by rubbing them with salt and coconut oil which further improves their appearance. A kilogram of this produce fetches nearly INR 30 to the farmer.

Apart from 10.6% hydroxy-citric acid and 15% reducing sugars in the rind, the seeds yield an edible fat (31%) also.

Problems and Prospects of *Kokam* Cultivation

The following are some of the most important problems of *Kokam* cultivation and its prospects:

1. Pressure on land
2. Reduction of the size of the holding
3. Popularization of tree species, coconut, arecanut and cocoa, in homestead gardening have, over the years, pushed *Garcinia* and many other useful plants to the background.
4. Variations in the yield
5. The dense shade of the foliage
6. The farmer's belief that it is a heavy feeder of plant nutrients (both inherent in soil and applied), tends to make this crop a liability

Against the background of all of the above problems, *Garcinia* with its tolerance to fluctuating water table and even prolonged inundation (waterlogging due to excessive rainfall, especially due to the south-west monsoon) is an excellent crop for the low-lying reclaimed soils, where farmers have only limited choice of remunerative perennial crops.

Kerala State in India is blessed with a vast diversity of *Garcinia* population, in particular, along the back-water and river-belt of the State. Exploiting this diversity with the ultimate aim of developing an ideal plant type with short stature, year-round fruiting and desirable fruit characteristics like higher acid edible pulp and seed looseness will bring about a radical change in *Garcinia* culture.

Trade Varieties

There are several trade varieties of *Kokam* available in Indian markets. The following are some of them:

1. Plain *Kokam*
2. Salted *Kokam*
3. Lonawala *Kokam*
4. Pakali *Kokam*
5. Khane (edible) *Kokam*
6. Khoba *Kokam*

Post Harvest Technology

Grading/Standards

No standards have yet been laid down for *Kokam* under the Agmark, ISI or PFA regulations. This demands early attention by Policy makers.

Table 3 The composition of fresh *Kokam*

Property	Content
Moisture	30
Protein (N: 6.25)	1.92
Crude Fiber	14.28
Total Ash	2.57
Tannins (polyphenols)	2.85
Pectin	5.71
Starch	1.0
Crude Fat (hexane extract)	10.0
Total acidity (as hydroxy-citric acid)	22.8
Pigment (total anthocyanins)	2.40
Ascorbic acid	60 mg/100 g (0.06%)
Carbohydrates (By difference, excluding Starch and Pectin)	36.4

Note: All values are in percentages on "moisture-free basis" and average of 2 analyses

Composition

The following Table 3 details properties of fresh *Kokam.*

Kokam seeds (comprising 8–10% of the whole fruit) contain about 25–30% fat which is greasy and whitish-yellow in color. The physico-chemical characteristics and composition of butter are given by Jamieson (1943) in great details.

Packaging and Storage

Kokam is packed in gunny bags. The shelf-life of fresh fruits is about 5 days at room temperature. Hence, sun-drying is practiced for the longer periods of storage/preservation.

Processed Products

For sun drying of *Kokam* rind, the fresh fruits are cut into halves and the fleshy portion containing seeds (8–10%) is extracted. The rind constitutes about 50–55% of the whole fruit. The rind contains about 15% acid (hydroxy citric acid). The *Kokam* of commerce is prepared by sun-drying the rind (skin) of the ripe fruits after repeatedly soaking it in the juice of the pulp. The product so dried constitutes the unsalted *Kokam* of commerce. Sometimes, after treatment with common salt during soaking and drying, the salted *Kokam* is produced. About 6–8 days are required for complete sun-dying.

Kokam seed is a good source of fat (about 25–30%), which is known as "*Kokam* Butter" in commerce, which is described below.

Kokam Butter

The seeds of the fruit yield about (25–30%) fat on the basis of weight of seeds or about 45% on the basis of kernels. It is a valuable edible fat known in commerce as "*Kokam Butter*". It is extracted mostly as a cottage industry by crushing the kernels, boiling the pulp in water and skimming off the from the top; or by churning the crushed pulp with water. Nowadays, oil is obtained by solvent extraction of seed kernels (after decortication of the seeds) in mechanical and motorized decorticators.

The characteristics of the *Kokam* Butter are given in the following table (Table 4).

Stearic Acid from *Kokam* Butter

Kokam Butter, like other *Garcinia* fats, is rich in combined stearic and oleic acids. It contains about 75% of mono-oleo-disaturated glycerides.

A method has been developed for the production of stearic acid from the fat with a yield of 45.7%. It possesses properties similar to piney tallow obtained from *Vateria indica* and may be employed in the sizing of the cotton yarn.

Kokam Syrup

Kokam Syrup is prepared from *Kokam* pulp/juice by mixing it with sugar and water and is used in the preparation of pharmaceutical products.

Table 4 The characteristics of *Kokam* Butter

Property	Content
Refractive index at 40 °C	1.4575
Saponification Value	187–191.7
Iodine Value	25–36
Unsaponification matter	2.3%
RM Value	0.1–1.0
Melting Point	40–43 °C

Kokam Concentrate

A process has been standardized at CFTRI, (Central Food Technological Research Institute) Mysore, India, for making *Kokam* concentrate, which, like Tamarind concentrate can be used as an acidulant in various preparations, such as curies, chutneys etc.

Kokam **Rind Acid** The isolation of the hydroxy citric acid (15%) in the pure form has been carried out on a small scale. Scaling up the process is in progress at the CFTRI. The pure acid finds a potential application in the preparation of pharmaceutical products.

Kokam Color

Kokam rind contains 2–3% anthocyanin pigments which constitute a promising source of natural color for acidic foods. The process for extraction of color and purification of the color pigment concentrate for coloring of foods and beverages has also been standardized by CFTRI, Mysore, India.

Ointment for Treating Carbuncles

An ointment made out of *Kokam* fat (butter), called, "white dammar resin"(exuded by *Vateria indica* tree) and wax is said to be effective in treating human carbuncles. This process is also standardized by CFTRI, Mysore, India.

A Fat-Soluble Yellow Pigment "Garcinol"

The pigment has been identified by chemical and spectral studies at the CFTRI, Mysore, India.

Gamboge

The commercial gamboge is a resin prepared from *Kokam* rind which was used extensively in the past in paints and varnishes.

An integrated process for manufacturing the above important products from *Kokam* is being developed at the CFTRI, Mysore, India, for the benefit of the industry.

Kokam fruit appears to be a promising industrial raw material for commercial exploitation in view of its important chemical constituents and industrial and domestic uses.

Uses

As Food Flavorant The fruit has an agreeable flavor and a sweetish acid taste. It is traditionally used in Konkan area of Maharashtra State, chiefly in the form of *Kokam*, prepared by drying the outer rind, soaking it repeatedly in the juice of the pulp and sun- during it. *Kokam* contains about 10% malic acid and a little of tartaric or citric acid. It contains 15% hydroxy – citric acid (Krishnamoorthy et al. 1982). It is used as a garnish to give an acid flavor to curries and also for preparing cooling syrups during hot months. *Kokam* is reported to have been imported earlier into Zanzibar from India during the 1960s. Besides Italy, and other foreign countries are also importing *Kokam* butter from India for use in confectionery.

In Medicine

The fruit is anthelmintic and cardiotonic and useful in treating piles, dysentery, tumors, body pains and cardiac complaints. A syrup from the fruit is given in bilious affections. The root is an astringent.

As *Kokam* Butter

Kokam butter is considered nutritive, demulcent, astringent, and emollient. It is suitable for ointment preparation, suppositories and other pharmaceutical purposes. It is used as a local application to ulcerations and fissures of lips, hands etc.

Kokam butter, as sold in the market, consists of egg-shaped lumps or cakes of light grey or yellowish in color with a greasy feel and bland oily taste. It is used mainly as an edible fat; it is also used as an adulterant of ghee (Indian native butter). As ordinarily met with, it contains seed particles as impurities. Refined and deodorized fat is white and compares favorably with high-class hydrogenated fats. *Kokam* butter is suitable for use as confectionery butter. However, as it solidifies with a rough surface, addition or blending with another fat is necessary to correct the defect. It is also suitable for candle and soap manufacture.

Anti-oxidant Property of Rind and Chemical Constituents

Recent investigations at the CFTRI, Mysore, India, have shown that its hexane extract incorporated at 0.2% level in groundnut oil showed a peroxide value of 2 meq/kg against 14 for TBHQ (0.02%), and 600 for the control treatment of groundnut oil. Thus, it possesses a good anti-oxidant property. Garcinol, a yellow-colored, fat-soluble pigment isolated from hexane-extract of the rind, however, did not possess any anti-oxidant activity level. Iso-garcinol- an isomer of garcinol showed only weak anti-oxidant activity.

The dried rind of *Kokam* (*Garcinia indica*) fruit contains 3 important constituents, namely, anthocyanin pigments, hydroxy-citric acid and garcinol. Garcinol, polyisophenyl phenolic pigment, is present to the extent of 2–3% in the dried rind of *Kokam*.

Application of Garcinol as a Coolant and Its Estimation

There is considerable variation in the color of *Kokam* butter sold in the market. The natural color varies from white to ivory-white to light-yellow, depending on its carotenoid content. In the dairy industry, where butter is produced on a large scale, the color varies from batch to batch. In order to get a uniform color, a small amount of a colorant like annatto or turmeric or a combination of both is added to the *Kokam* butter.

Garcinol, when added at 0.3% level, imparted an acceptable yellow color to *Kokam* butter. Its retention was 98.4% after 90 days storage at 4 °C and was 99% after 15 days storage at 25 °C. On conversion of butter into ghee (Indian native butter), by heating, 94% of the added garcinol was retained (Krishnamurthy and Sampathu 1988). Besides, the method of estimation of Garcinol in *Kokam* butter, a method has also been developed by the same authors to determine the same in ghee (Indian native butter) at CFTRI, Mysore, India.

References

DMI, Government of India, New Delhi Report (1968) Directorate of Marketing and Inspection, Ministry of Agriculture, Government of India. 1968. Marketing of Minor Spices in India. AMA 163 Report, Manager of Publications, New Delhi, India

Jamieson GS (1943) Vegetable fats and oils, 2nd edn. Reinhold Publishing Corporation, New York, p 87

Krishnamurthy N, Sampathu SR (1988) Anti-oxidant properties of kokam rind. J Food Sci Technol 25(1):44–45

Krishnamurthy N, Lewis YS, Ravindranath B (1982) Chemical constituents of kokam fruit rind. J Food Sci Technol 19(3):97–100

Muralidharan HG (1976) Raw material survey of resources and newer sources of fat and oils- "*Kokam*". In: Proceeding of the symposium AFST/OTAI/CFTRI, Mysore, India, June 1976

Naik KC (1949) South Indian fruits and their culture, 1st edn

Spices Board, Government of India (1988) Annual statement of exports of spices and spice products from India. Spices Board, Government of India

Juniper

Abstract The chapter at first discusses description and distribution details of the plant. Subsequently crop management aspects would be discussed. The chapter will discuss, also, at length, post harvest technology with a special emphasis on Juniper berries and Juniper volatile oil.

Keywords Description · Distribution · Berry · Volatile oil · Alcoholic beverages

Botanical name: *Juniperus comm*unis Linn.
Family: Pinaceae

Description and Distribution

The common Juniper, which is found in the Himalayas, from Kumaon westward, at an altitude of 1520–4270 meters, is an evergreen shrub, sometimes attaining the height of a small tree up to 3 meters, with an erect trunk and spreading branches, covered with a shreddy bark. The trunk diameter is 25–30 cm. The leaves are straight and rigid, oval-shaped, about 6–13 mm long, and with sharp prickly points. It is a dioecious; the male and female flowers produced in April and May are usually borne on separate plants; the male flowers in short catkins and female flowers in small cones. The fleshy berry-like fruit does not ripe until the second year after planting. The roundish fruit is sub-globose, bluish black, dark-purple, when ripe, 10–13 mm in diameter, covered with a waxy look; the three scales comprising the fruit occasionally gaping and exposing seeds; bony seeds, usually 3, elongated, ovoid, embedded in pulp.

The plant is very variable with a number of geographical varieties and garden forms; it becomes prostrate, not more than 60–90 cm in height, at higher altitudes in the Himalayas. It grows wild in many parts of Europe and Asia, ranging as far east as the Western Himalayan mountains in North-Eastern Asia. In North America, the

K. P. Nair, *Herbal and Acidulant Tree Spices*,
https://doi.org/10.1007/978-3-031-04360-4_8

bush or tree occurs south to Pennsylvania State and west to Illinois State, New Mexico and northern California States.

Area, Production and Exports

No reliable statistics are yet available on the above-mentioned aspects, since Juniper is not grown in India on a commercial scale.

Crop Management

Soil and Climate The quality, appearance, flavor and essential oil content of the berries of Juniper depend to a great extent upon the geographical origin, and local conditions, such as, climate, soil, altitude, and, exposure to the sun. Trees at elevated altitudes, and in warm sunny places, produce the best berries. Those from Italy have always been considered to be of the best quality, following next are the Himalayan berries. Juniper thrives best on sunny heights and mountain slopes, exposed to warm sun.

Propagation

The propagation is usually by seeds under natural conditions, the seeds are dispersed by the birds, who devour the fruits.

Flowering and Fruiting

The plant flowers in March-April and the fruits ripen in August-September of the second year since planting

Harvesting of Berries

Ripe (blue colored) berries yield more essential oil than green berries. Therefore, only ripe berries need to be harvested as follows:

Juniper berries are collected in a simple manner. Harvesters carefully thrash branches of the bush with wooden sticks, until ripe (2 years old) berries drop into collection baskets held below the bush, or, polythene/cloth sheets spread below the bush. Green (1 year old) berries remain on branches which can be collected the following year.

Post-harvest Technology

Drying of Berries

Strictly speaking, they are fruits and not berries, technically, (botanically), but, they look like berries. The freshly harvested material is transported to the nearest village, where berries are sifted, winnowed, and spread out in a thin layer on the floor for drying under the sun or in a covered shed. Care must be taken to prevent fermentation, which takes place easily if the fresh berries are stored in sacks. Any berries damaged by fermentation are sold at a discount, because, they possess an inferior flavor and poor appearance.

Grading

The quality of berries and essential oil manufactured therefrom, varies with the region where Juniper is grown. The quality of the Italian, Hungarian, Czechoslovakian, Yugoslavian, and other European berries are ably described by Guenther (1972). They are potential producers of Juniper berries. The Italian berries are considered the best, followed by the Hungarian berries. In Italy, Juniper grows specially well at altitudes of 600–1200 meters above mean sea level. Italy is, by far, the largest producer of Juniper berries. The berries are large, deep blue in color, rich in essential oils, and excellent in flavor. Harvesting is usually done from August 20 to September 10. The total production per annum varies greatly depending on the demand. In normal years, it averages between 3500–4000 tonnes. Prior to the World War II, the bulk of the quantity produced was exported to Germany (Anon 1939).

Indian Grades

The Italian berries are usually offered in four grades, as follows:

Grade 1: Large, hand-picked berries, sold to the trade in cellophane-wrapped, 1 lb lots, and retailed through groceries to households for cooking and domestic use.

Grade 2: The next smaller size, graded in machines. Sold in sacks for general kitchen use, in particular, for making of preserves.

Grade 3: Also, machine-graded. These serve various industrial purposes; for flavoring of sauces, making of preserves, and for distillation of alcoholic beverages. Their price averages one-half of the Grade 1 berries. Occasionally, this grade of berries is sold in powdered form.

Grade 4: The smallest of berries, machine-graded, primarily for the distillation of oil.

The Juniper berries of commerce are obtained by drying of ripe fruits. The oil of Juniper is obtained by steam-distillation of these very dried berries and is in a great demand (Guenther 1972).

Composition of Juniper Berries/Fruits

Fruits contain, besides the volatile oil, fermentable sugars (33%), resin (8%), juniperin, probably a mixture of tannin and sugars (0.36%), fixed oil, protein, wax, gum, pectins, organic acids (fumaric, acetic, malic, oxalic and glycolic) and potassium salt. They are a good source of ascorbic acid also (about 35 mg/100 g).

Juniper fruits have gin-like aroma and a sweet terebinthinate taste with a somewhat bitter after-taste (Pruthi 1998).

Adulteration

Formerly, juniper berries were frequently found adulterated with berries of inferior varieties/wild species, namely, *Juniper communis* var. "nana" Loud., *Juniperus oxyeedrus* Linn., but, due to the following reasons, such as, (1) present high cost of collection of berries from wild species /varieties from higher altitudes and (2) stricter enforcement of quality standards, they are no longer admixed in true Juniper berries now-a-days.

Juniper Berry Essential Oil (Volatile Oil)

Juniper oil is obtained by steam distillation of ripe fruits, the yield varies according to the quality of the fruits, distilled. The berries contain 0.5–2.0% essential oil. Ripe (blue colored) berries yield more essential oil than unripe (green colored) ones. The quality and yield of oil depends greatly upon the geographical origin of the berries. The oil from green unripe fruits is inferior. In over-ripe berries the oil changes into resin. The bulk of commercial oil is obtained as a by product during distillation of alcoholic beverages. The oil, however, is partly deprived of natural oxygenated odoriferous compounds.

Composition of Juniper Volatile Oil

Juniper oil is colorless or pale-greenish yellow limpid liquid with a characteristic odor of the fruit and with a somewhat burning bitter flavor. On storage, the oil turns viscous and acquires a turpentine odor. The characteristics of fresh oil obtained by steam distillation of the ripe fruits usually vary within the following limits:

Specific gravity at 15 °C: 0.867–0.882
Refractive Index at 20 °C:1.472–1.484
Optical Rotation up to – 13 °C occasionally dextro – rotatory
Acid Value: Up to 3
Ester Value: 1–12
Ester value after acetylation: 19–31
Solubility at 90% alcohol, 1 in 5–10 volume becoming less soluble with age

The fruits obtained from Hoshiarpur district in Punjab State, India showed the following characteristics:

Oil yield 0.83%
Specific gravity at 27 °C: 0.918
Refractive Index at 25 °C: 1.482
Optical rotation: +20.8 °C
Acid value: 4.7
Ester value: 20.5

The I.P.C. requirements for the oil are as follows:

Specific gravity at 20 °C: 0.862–0.892
Refractive Index at 20 °C: 1.476–1.484
Optical Rotation: 1–15 °C (Pruthi 1987)

The oil contains α-pinene as the major constituent, together with smaller amounts of camphene, cadinene, Juniper camphor (possibly a sesquiterpene alcohol), a hydrocarbon (Junene) with strong diuretic properties, terpinenol, certain unidentified oxygenated compounds possessing characterristic Juniper color, and traces of esters (Guenther 1972).

Manufacturing Alcoholic Beverages

In major Juniper producing European countries, such as, Hungary, Czechoslovakia and Yugoslavia, large quantities of Juniper berries are used domestically for fermentation and subsequent distillation of popular alcoholic beverages, such as, "Borovicka", which is a type of gin. In Germany, Austria and Switzerland, the same drink is called "Steinhager". This spirit is nothing but an alcoholic distillate of fermented Juniper berries containing as much as 40–50% or more of alcohol. Slovaks are, in particular, fond of this strongly flavored beverage.

Uses

As Food and Food Flavorant

Fruits are employed to flavor gin and food products. They are sometimes used as an article of food. Large quantities of fruits are used in Europe to prepare alcoholic beverages of the nature of gin. For this, the fruits are crushed, immersed in warm water and fermented; the fermented mass is then distilled and rectified. 1000 kg of fruits yield 16–18 liters of beverage, containing 40–55% alcohol and 5–6 kg of volatile oil.

Juniper Oil

It is largely used in compounded gin flavors, liquors and cordials. A twice rectified oil has high flavor value. Imitation – Juniper oils have been produced.

Other Parts of the Juniper Tree

All parts of the Juniper tree contain a volatile oil. A terebinthinate juice exudes from the tree and hardens on the bark; it has erroneously been considered identical to gum sandarac, obtained from *Tetraclinis articulata*.

Terminal twigs and needles yield (0.15–0.18%) of a bright yellow-colored oil (specific gravity at 20 °C of 0.8531), with the characteristic odor of Juniper oil; it contains d-α-pinene, camphene and cadinene.

In Medicine

The fruits and volatile oil possess carminative, stimulant and diuretic properties. They are useful in remedying different forms of dropsies, especially, in conjunction with other drugs. They have been used to correct disorders of urogenital tract, such as gonorrhoea, gleet and leucorrhoea, and, in certain cutaneous diseases. Dried fruits are sold in Indian markets (North Indian) and are reported to be imported from Nepal via Patna (in Bihar State, India) and are used in the treatment of gonorrhoea.

Volatile Oil

The oil of Juniper berries has been, for long, used as a good diuretic. The oil should be used with much caution because of its irritating effects on inflamed organs.

Juniper Wood

The wood has diuretic properties, sudorific and, a blood purifier. It is employed in the treatment of gout, rheumatism and cutaneous diseases.

Other Uses/By Products

Exhausted Fruits/Berries

The exhausted fruits (left over after distillation of oil) on concentration and repeated extraction with warm water yield 30–38% – a by product known as "*Succus Juniperi*". The preparation, consisting chiefly of inverted sugar, was formerly used in Europe as a diuretic and sudorific. Exhausted fruits are used as a feed for live-stock. Its composition is given in the following table (Table 1).

Juniper Needles

Juniper needles are rich in ascorbic acid (88 mg/100 g); they also contain resin, wax and esters.

Table 1 Composition of exhausted fruit of Juniper

Property	Content
Moisture	23.72%
Crude protein	6.23%
Ether extract	10.75%
Crude fiber	27.16%
N-free extract	38.0%
Ash	4.14%

Note: The Ash is rich in Calcium and Potassium. Feeding trials on sheep showed the following digestibility coefficients:
N-free extraction 66%
Protein 39%
Ether extract 37%
Crude fiber 20%

Roots

The fruits and roots yield brown and purple dyes respectively. In Russia the bark is used for tanning leather.

Wood

The wood of Juniper (w.331/cft) is brownish, moderately hard, durable, fragrant, highly resinous and easy to season. It is usually available in small sizes and used for fencing, veneering, turnery and as a fuel. The wood and young twigs are burnt as incense also.

Juniper Wood Oil

It is obtained by steam distillation of wood and has the following characteristics:

Density at 15 °C: 0.8692
Optical Rotation: 21.03°
Refractive Index at 20 °C: 1.4711
Acid Value; 0.9
Ester Value: 6.7
Solubility in 90% alcohol: 1 in 7 volume or more, with slight turbidity

It contains candinene and a sesquiterpene. Commercial Juniper wood oil is usually obtained by distilling turpentine with Juniper wood and twig. More often, it is a mixture of turpentine and Juniper oil. The bark oil obtained by steam distillation of bark (yield 0.25–0.50%), contains juniperene, juniperol, α-pinene and silvestrene (Guenther 1972).

References

Anon (1939) Ber. Schimmel & Co. p 88
Guenther E (1972) The Essential Oils. 5th reprint. Juniper berry oil. pp 370–381
Pruthi JS (ed) (1987) Bibliography on 57 Spices and Condiments for the years 1972–86. AFST (1), Delhi, India. Chapter, pp 1–168
Pruthi JS (1998) Mace and nutmeg. In: Spices and condiments. National Book Trust, New Delhi, India. pp 151–55, 167–72

Hyssop

Abstract The chapter at first will discuss description and distribution. Further, it would discuss area, production, exports, crop management, processed products and end uses.

Keywords Description · Distribution · Processed products · Food flavorant

Botanical name: *Hyssopus officinalis* Linn.
Family: Labiate

Description and Distribution

Hyssop is an aromatic perennial shrub, 30–60 cm high, found in the Himalayas, from Kashmir to Kumaon at altitudes of 2435–3335 meters above mean sea level. It is a native of southern Europe and temperate zones of Asia. It has been naturalized in the USA where it grows in gardens along the roadside. Branches are erect or diffuse, leaves are sessile, linear oblong or lanceolate, obtuse, entire; flowers are bluish purple, in auxiliary tufts arranged unilaterally on terminal branches; nutlets are dark brown, narrow, trigonous, smooth. Leaves and flower tops constitute condiment or spice.

The herb grows wild in countries bordering the Mediterranean. It is cultivated in Europe, particularly in southern France, mainly for its essential oil. It is also grown in gardens for ornamental purpose.

In India, it has been successfully cultivated at Baramulla (altitude 1675 meters above mean sea level) in Kashmir.

K. P. Nair, *Herbal and Acidulant Tree Spices*,
https://doi.org/10.1007/978-3-031-04360-4_9

Crop Management

Soil and Climate The plant thrives in light rich soil in hill stations. It may be propagated by seeds, cuttings and divisions. Like the Lavender plant Hyssop grows in calcareous soil (Guenther 1973).

Planting and Cultivation Planting of Hyssop resembles that of Lavender. Slips are grown in nurseries during fall, and in the following spring, they are transplanted in the field at 1 m distance in rows which are 1 m apart from each other.

Harvest The maximum yield of oil is obtained right after the bloom. The properties of the oil vary only little during the flowering period and withering, and oil in all the above ground parts of the plant is of uniform quality (Chiris 1929). First harvest can be done in July, two years after planting.

Yield One hectare with approximately 10,000 plants yield, on an average, 3000 kg of plant material. A field of Hyssop has a life span of 6–7 years.

Post Harvest Technology

The harvested plants are uniformly dried in sheds to "clover-dry" condition.

Composition The herb contains volatile oil, fat, sugar, choline, tannins, carotene (108.1 mg/100 g) and xanthophyll (355.6 mg/100 g). The tops contain ursolic acid (0.49%). A glucoside, diosmin, which on hydrolysis yields rhamnose, glucose, and a glucone, diosmetin (4- methyl luteolin), which has also been isolated. Hyssop is stated to yield a greyish-green dye. The fresh herb contains iodine in a concentration of 14 mg/kg.

Processing Technology

Distillation of Essential Oil Most of the Hyssop originates from the plantings; the wild growing kinds being so scattered that collection becomes too expensive. Wild plants are, therefore, used only when the cultivated material does not meet the demand.

Distillation of Hyssop is carried out by farmer producer in the type of stills usually employed for Lavender. Some of these are of the old-fashioned direct-fire-stills, provided with a grid, above which the plants are charged. Distillation of one charge lasts several hours.

Yield of Essential Oil The oil yield varies greatly with the condition of the plant material, depending chiefly upon its dryness. It averages from 0.15–0.3% for fresh to clover-dry, and from 0.3–0.8% for clover-dry to completely dried material. Maximum yield is obtained from plants harvested just after the opening of the blossoms. The oil obtained from all the above round parts is of uniform quality.

Physico-Chemical Properties of Essential Oil Hyssop oil is colorless or greenish yellow with an agreeable aromatic, somewhat comprehensive odor, and warm, slightly bitterish taste. It has the following range of constants:

Specific gravity at 15°: 0.935–0.952
Optical rotation: 15.75^0–18.3^0
Refractive Index at 20 °C: 1.4783–1.4829
Saponification value: 2.812.1

It is soluble in 1 vol of 80% alcohol, clear to cloudy with more; in exceptional cases, incompletely soluble in 80% alcohol. Its quality is more or less uniform, irrespective of the part from which the oil has been obtained; the properties vary slightly when the oil is distilled from herbs harvested during the flowering and withering periods.

Steam-distilled oil from the herb collected from Baramulla (yield 0.36% and 0.7% respectively from fresh and dried herb) had the following physical constants:

Specific gravity at 15 °C: 0.9375
Refractive index at 20 °C: 1.4778
About 50% of oil consists of 1-pinocamphone.

Adulteration Because of its relatively high price, Hyssop oil is occasionally adulterated with spike, Lavendar or Rosemary oils.

Most dangerous are the additions of camphor oil fractions, because they are not easily detected in routine analysis. In fact, the presence of camphor may denote a high content of 1-pinocamphone, the latter in pure oil is indicative of a good quality. Obviously, the addition of d-camphor would affect levorotation of pure Hyssop oil, which helps in detection of adulteration.

Uses

As Condiment The leaves and flowering tops of Hyssop have an agreeable aromatic odor and a warm, pungent, bitterish taste. They are employed as flavoring ingredients for salads and soups, and also, in the preparation of liquors, bitters, tonics, and specially in liquors of French make, in imitations of Chartreuse and Benedictine etc. and also in perfumes. Occasionally, the green parts are used as pot-herb.

In Medicine

Hyssop is a stimulant, carminative and pectoral, used in colds, coughs, consumption and lung complaints. An infusion or tea prepared from the plant is said to be effective in nervous disorders, toothache, and in pulmonary, digestive, uterine, and urinary problems. Leaves are stimulating, stomachic, carminative and useful in treating hysteria and colic. The leaf juice is employed in the expulsion of round worms. The crushed herb is applied as a resolvent and vulnerary.

Infusion of plant is reported to be useful in asthma and coughs. Steeped in hot water, it is used as a resolvent and vulnerary, as fomentation to cure wounds, sprains and strains, muscular rheumatism and to clear discoloration of the skin due to blows. It is also used as a salve in catarrhal opthalmia.

Hyssop was at one time official in some pharmacopeia in Europe. "*Zufah yabis*", a drug sold in Indian bazaars and imported from Syria and Iran, has medicinal properties almost similar to those of Hyssop. It has been referred to as *Hyssop officinalis* by some authors, but, this identity is, indeed, doubtful.

Uses of Hyssop Volatile Oil

1. Flavoring beverages: Hyssop oil is used as a flavoring agent in bitters and tonics, especially in French liqueurs of the Chartreuse and Benedictine type.
2. Oil in Perfumery and Cosmetics: It is also used to some extent in perfumes with a spicy odor
3. In Medicine: In small dose (1 or 2 drops), the oil promotes expectoration in bronchial catarrh and asthma. It produces mild colonic convulsions when injected intravenously. In dogs, intravenous injection of 1–2 ml/kg of a saturated solution of essential oil in 33% ethyl alcohol causes a little increase in blood pressure and respiration, then mild colonic convulsions, then decrease in blood pressure and increase in heart rhythm.

References

Chiris (1929) Perfums France 7, 201

Guenther E (1973) Oil of hyssop. In: The essential oils, 5th Reprint. D van Nostrand Co. Inc., New York

Marjoran

Abstract The chapter will begin with a description and distribution of the plant. Further, it would discuss area, production, exports, crop management, post harvest technology, end uses, especially as a food flavorant and use in perfumery.

Keywords Description · Distribution · Food flavorant · Medicine · Volatile oil · Perfumery · Cosmetics

Botanical name: *Marjorana hortensis* Moench.
Family name: Labiatae

Description and Distribution

Marjoram is a perennial aromatic herb, which is native to Southern Europe, North America and Asia Minor. It is the dried leaves of marjoram or sweet marjoram with or without flowering tops in small proportion, which constitute the leafy spice of commerce. It is an aromatic herb of the mint family, 30–40 cm high. It is extensively cultivated in India. Sweet Marjoram is characterized by a strong spicy and pleasant odor. The flavor is fragrant, spicy, slightly sharp bitterish and camphoraceous. Though a perennial, it is treated as an annual under cultivation. A native of Southern Europe, it grows in Western Asia, South and North America, France, Germany, Hungary, Greece, Romania, Spain, Portugal, USA, England, the Mediterranean and North Africa. It is the first cousin of increasingly popular Oregano, belonging to the same mint family. The color of the dried herb is light green with a slight greyish tint. The whole leaves are grey-green in color, small, with hairs on both sides of the leaf. When examined under a microscope of low power, many dot-sized oil glands are visible on the leaf. They yield 3.5% oil.

K. P. Nair, *Herbal and Acidulant Tree Spices*,
https://doi.org/10.1007/978-3-031-04360-4_10

Area, Production and Exports

No data are available for India, though world trade in Marjoram as early as 1977 amounted to 408 tonnes valued at US $ 521,000. Until now, India has not entered world trade in Marjoram, but, the crop holds huge export potential.

Crop Management

It grows in any well-drained, fertile garden-loam soil. Marjoram, though a perennial, is treated as an annual under cultivation.

Propagation

Marjoram is propagated by seeds and cuttings. Seeds are sown in October in plains and in hills from March to the middle of June. Seeds are sown in pots and the seedlings, when large-enough to handle, are transplanted in the field 20–25 cm apart in rows which are spaced 30 cm apart. Propagation by cuttings is sometimes done at higher elevations.

Influence of Growth Regulators on Growth, Herbage and Essential Oil Yields

Investigations on the influence of 3 growth regulators, namely, Gibberlic Acid (GA), Cycocel and Triacontanol, on Marjoram revealed that application of GA at 200 ppm increased fresh herb recovery and recorded maximum oil yield. Cycocel at 500 ppm and Triacontanol at 6 ppm increased recovery of the herb oil yield over control treatment, but, were next to GA application (Vasundhara et al. 1992).

Harvest

Effect of N and P on growth, herb yield and essential oil content.

Results of the investigation on 3 harvests of Marjoram done during one year revealed that application of the highest dose of N at the rate of 320 kg/ha gave significantly higher herbage yield (11.54 tonnes/ha) and oil yield of 52.79 kg/ha. Though the application of P at 40 kg/ha recorded the highest herb yield (10.01 tonnes/ha) and that at 120 kg/ha gave the highest oil yield of 48.87 kg/ha, they were found statistically on par with P application at 80 kg/ha, which gave 9.85 tonnes/ha of herbage yield and 46.84 kg of oil yield (Farooqui et al. 1994).

Marjoram has a wide range of variability in various yield-attributing traits offering ample scope for selection. The total herbage yield was positively correlated with plant height, plant spread, fresh weight of herb, dry weight of spike and total dry weight of whole plant. However, the oil yield was found to be non significantly correlated with all the characters investigated. Prasanna Kumar et al. (1994a, b) also investigated the floral biology of Marjoram.

The crop is ready to harvest in about three and a half months from the date of planting. The tops are cut when the plants are near flowering and dried in the shade. The volatile oil content is maximum when the plant is harvested before seed formation.

Flowering

The marjoram plant commences flowering 60 days after planting, which continues for one more month. Flowers are bracteate, with calyx (5 sepals) and corolla (5 petals); are sessile, complete, bisexual, spathulate, zygomorphic, gamopetalous, bilipped; aestivation is imbricate, androecium 4, ovary is bicarpellary, syncarpous. Anthesis occur between 8 A.M. and 3 P.M. Anther dehiscence takes place from 10 A.M. to 4 P.M. The maximum receptivity of stigma was found on the day of anthesis. Pollen grains were found to be dull white in color, oblong to oval shape, with 43.74 μ in length and 27.70 μ in breadth (Prasanna Kumar et al. 1994b).

Post Harvest Technology

Drying, Packing and Storage

Leaves/tops are dried in the shade and packed in containers and closed shut.

Composition

Analysis of the dry herb showed the following composition:

Moisture: 7%
Protein: 14.31%
Fixed oil: 5.60%
Volatile oil: 1.72%
Pentosans: 7.68%
Fiber: 22.06%

Ash: 9.69%
Tannin, an astringent substance, and ursolic acid (0.21% in tops, 0.05% in stem) are present

There is wide variation in the mineral content of Indian, French and German sweet Marjoram herbs with the following composition:

Total ash: 6.3–24%
Sand: 0.66–14%
Sand-free ash: 5.4–14.3%
Potash: 18.3–20.2%
Sodium: 0.65–0.68%
Calcium: 17.60–24.8%
Phosphorus: 8.9–9.1%
Iron: 6.1–7.3%
Silica: 19.4–26.5%
Magnesium: 4.8–6.7%
Manganese: Traces to 1.05%
Chlorine: 1.51–2.05%

Commercial sample of Marjoram should contain the following:

Crude fiber: 22% (maximum)
Total ash: 13% (maximum)
Acid-insoluble ash: 3.5% (maximum)
Volatile oil: 1% (minimum)
Moisture: 10% (maximum)
Total ether extract: 6.5% (minimum)
Sweet Marjoram contains about 14.5% protein, 5.6% fixed oil and 7.68% pentosans

Volatile Oil

Steam-distillation of the leaves and flowering heads yield a volatile oil, known to the trade as "Oil of Sweet Marjoram" (yield from fresh flowering herb: 0.3–0.4%; dry herb: 0.7–3.5%). The oil is colorless or pale yellow to yellowish-green, with a tenacious odor, reminiscent of nutmeg or mint.

A sample of Indian oil (from Delhi) had the following physico-chemical properties:

Density at 15 °C: 0.9346
Refractive Index: 1.5062
Optical rotation: +40.25^{0}
Acid value: 4.8
Saponification Value: 8.32
Saponification Value (after acetylation): 128.4

Phenol content: 47.7%
The oil consisted of carvacrol (36.4%), eugenol (6.7%), chavicol (4.6%), d-linalool (30.6%), methyl-chavicol (3.2%), d-terpinol (4.8%) and caryophyllene (7.6%)

Oils of European origin differ considerably from the Indian oil and have usually the following ranges of characteristics:

Density at 15 °C: 0.894–0.901
Refractive Index: 1.470–1.476
Optical rotation: +15 °C to 25 °C
Acid value: up to 1.5
Ester value: 10–38
Ester value (after acetylation): 41–78; soluble in 1–2 vols and more of 80% alcohol. They contain about 40% terpenes (mainly terpinene) but are free from phenols; d-terpineol and terpinenol are also present
Sweet Marjoram oil is often confused with thyme oil (from *Thymus* spp.) and origanum oil (from *Origanum* spp.)

Uses

As Food Flavorant

Marjoram leaves are used by the industrial manufacturers for flavoring liver and Polish sausages and cheese and also in soups, stews, dressings, salads, egg and vegetable dishes (peas, beans), cheese, fancy meat (lamb ad mutton) dishes and sausages and poultry dressing. The leaves of the plant are used fresh or dried and are highly esteemed as a condiment for seasoning foods; they are used also as a poultry-seasoner. Fresh leaves are employed as garnish and incorporated in salads, they are used also for flavoring vinegar. Dried flowering tops are used for sachets and potpourri. The aromatic seeds are used in confectionery and French confitures.

In Perfumery and Cosmetics

The oil is employed to a small extent in high grade flavor preparations and perfumes, and in soap and liquor industries.

In Medicine

Sweet Marjoram is considered a carminative, an expectorant, and a tonic; leaves and seeds are astringent. An infusion of the plant is used as a stimulant, surorific, emmenagogue, and galactagogue; it is reported to be useful in asthma, hysteria and paralysis.

It is used in Homeopathic mother tincture. It acts upon the generation of organs, in particular, in women. It is also used successfully for troubles arising from sexual irritation.

It's oil is used as an external application for spasms, bruises, stiff and paralytic limbs and tooth ache, and for the fomentation in acute diarrhoea. Intravenous injection of 1 ml/kg of body weight of dogs of a saturated solution of essential oil in 33% ethyl alcohol increased peristaltic movement of the intestine.

Leaves and seeds of Marjoram are astringent and are reported to provide a remedy for colic.

References

Farooqui AA, Deviah KA, Shridharayya M, Vasundhara M (1994) Effect of nitrogen and phosphorus on growth, yield and essential oil content of marjoram. Indian Perfum 38(1):9–14

Prasanna Kumar GH, Khan MM, Farooqui, AA (1994a) Variability and correlation studies in marjoram. Indian Perfumer 38(1):1–4

Praanna Kumar GH, Khan MM, Farooqui AA, Joshi SS (1994b) Studies on floral biology of marjoram. Indian Perfumer 38(1):5–8

Vasundhara M, Farooqui AA, Deviah KA, Shridharayya M (1992) Influence of some growth regulators on the growth, herbage, and oil yield in marjoram (*M. hortensis* Moench.). Indian Perfum 36(3):171–174

Japanese Mint

Abstract At first, the chapter will discuss description and distribution of the plant. Subsequently the chapter will discuss area, production, exports, crop management, plant protection. There would be an elaborate discussion on mint oil production.

Keywords Description · Distribution · End products · Essential oil · Menthol

Botanical Name: *Mentha arvensis* Linn.
Family: Labiatae

Description and Distribution

Mints are a group of plants belonging to family Labiatae, which yield essential oil on distillation. Commercial cultivation of mints in India, is of recent origin. It is used in "chutney" and as an old popular household remedy for relieving cold and cough. It belongs to genus *Mentha*, comprising about 40 species of aromatic, perennial herbs, distributed mostly in the northern hemisphere (Europe, America, Japan, China, Brazil and Formosa). In India, about 8 species of *Mentha* are reported to occur or grow. However, the world demand for peppermint oil and menthol is met from the following 3 species, approved/recognized by the ISO (International Organization for Standardization) for the purpose of International Quality Standards and international marketing.

1. Mint or Japanese mint (*Mentha arvensis* Linn.) subsp. *haplocalyx* Briquet, Holms
2. *M. piperita* Linn. var. *piperita*
3. *M. spicata* Linn. syn. *Mentha viridis* Linn.

Japanese mint, which is covered in this chapter, is a perennial herb with creeping rootstock and erect stem, is 1–2 quadrangulate branched, and, short with short dense hair; leaves are 2.5–5 cm long, shortly petiolated, oblong – ovate, or, lanceolate, or obtusely or acutely serrate; the upper leaves are similar and large, covered with minute hairs, which on the ventral side of the mid-vein are promi-

K. P. Nair, *Herbal and Acidulant Tree Spices*,
https://doi.org/10.1007/978-3-031-04360-4_11

nent. Flowers are in axillary distant whorls, none at the top; bracts are acute, shorter than flowers; calyx is hairy, throat is marked, teeth are triangular or lanceolate, fruiting calyx is scarcely enlarged and pink; corolla is white or creamy, hairy without or within tube, as long as calyx or slightly exserted, lobes are spreading; stamens are exserted; style exserted; nutlets are smooth and dry.

The plant grows to a height of 0.4–0.8 m, branching freely, beginning from its base. Flowers appear in May–June and again in September–November, under cultivation. Fertilization is by cross pollination and as such the seeds do not produce genetically pure type, besides being invariably sterile. During winter months, underground stems called stolons are formed at the end of each creeping rootstock. The crop, as a rule, is thus raised from these stolons in early spring moths; the runners, on the other hand, are produced by the plant above the ground during rainy season and are also used for propagation/raising the crop in August and September. It is a hardy plant which grows in a wide range of soils and climate under irrigated conditions. For raising commercial plantations, tropical to sub-temperate climate, elevation below 1000 m and light showers at the planting time are required by this species.

Area, Production and Exports

In 1952, rooted suckers of Japanese Mint plant (*Mentha arvensis* Linn. var. *piperascens*) were obtained from Japan and planted in Jammu and Kashmir in over 2000 acres and in the *Terai* region of Haldwani district in former Uttar Pradesh, now Uttarakhand, in India, covering more than 3000 acres. The other centers of production are Chakrohi in Jammu and Kashmir, Kuppam in Andhra Pradesh, and Jalandhar in Punjab State, all in India.

Mint cultivation is largely confined to North India because of the prevalent subtropical and temperate climate. Uttarakhand is contributing almost the entire production of *M. piperita, M. citrata* and most of the production of Japanese mint (*M. arvensis*).

No official statistics are available on the production of the various mint oils in India. According to one estimate, Japanese mint is cultivated in about 37,000 ha with a total production of about 5300 tonnes. The details of the various types of mints, which are cultivated in India are given in the following Table 1.

Table 1 Cultivated Mint species, their main constituents and projected production in India in 1993

Common name	Botanical name	Main constituents	Production (tonnes)
Japanese mint	*Mentha arvensis*	Menthol (70–80%)	6000
Peppermint	*Mentha piperita*	Menthol (50–65%)	50
Bergamot mint	*Mentha citrata*	Linalool (40–50%) and Linalyl acetate (35–40%)	50
Spearmint	*Mentha spicata*	Carvone (57–71%)	60

Source: (1) Indian Perfumer, Vol 37, 1993 (2) Kothari and Singh (1994)

Export of Mint Oils from India

India is an important country in mint oil trade, which is exported to many countries. The annual average growth rate of export of mint oils was remarkable: at 53% in quantity and 61% in value in the last 4 years leading up to 1992–93. Export of mint oils which was 158 tonnes in 1989–90 has shot up to 1273.2 tonnes in 1992–93. In terms of value, the increase has been from INR 329.83 lakhs in 1989–90 to INR 2282.18 lakhs in 1992–93. However, export was marginally less in quantity and value in 1993–94.

The internal demand for various mint oils is increasing rapidly owing to their diversified use in cosmetic, pharmaceutical, confectionery and liquor industries. Furthermore, the mint oils have large export potential owing to unlimited availability of skilled and unskilled labor, resulting in low cost of production (Kothari and Singh 1994). Being high in production, Japanese mint oil is exported in larger quantities. It was 112.8 tonnes in 1989–90 and 647.9 tonnes in 1993–94 (Spices Board 1994). More recent trend in the dramatic increase in the export of mint oil is summed up in the following Table 2. The 1996–97 export of mint oil reached an all-time high of 2120 tonnes, valued at INR 104.90 million as compared to just INR 40.6 million in 1995–96.

Crop Management

Climate and Soil

Rainfall, maximum and minimum temperatures, altitude, of the mint growing areas bear considerable influence on the growth of the plant and oil yield and free menthol. Details are described in the following Table 3.

Temperate to tropical climate is suited for good plant growth. Sunny weather with moderate rain is conducive to its luxuriant growth.

Subtropical climate is ideally suited for the successful cultivation of *Mentha arvensis*. A deep soil rich in humus which retains moisture, is quite suitable. Mentha can withstand waterlogging for a short period, but, as far as possible waterlogging should be avoided. Excess water should be drained or else the crop would be badly damaged.

Table 2 Recent jump in the Export of Mint oil from India

Year	Quantity (tonnes)	Value in INR (million)
1994–95	1583	435.7
1995–96	1242	405.8
1996–97	2120	1049.0

Table 3 Effect of rainfall, maximum and minimum temperatures and altitude on mint oil yield

State	Stations	Percentage oil yield from menthol (%)	Leaves (volume/weight, Free on moisture-free basis)
Jammu and Kashmir	(1) Jammu/Canal Road	3.4	72.9
	(2) Jammu/Chatha	3.2	74.6
	(3) Srinagar (Kashmir)	1.6	77.0
	(4) Katra (Jammu)	2.4	77.6
Tamil Nadu	(1) Chennai	3.0	82.0
	(2) Coimbatore	3.6	73.2
	(3) Nilgiris	2.4	52.2
Himachal Pradesh	Bushar	3.6	80.5
Uttarakhand	(1) Kanpur	2.2	72.7
	(2) *Terai* region	4.2	71.2
	(3) Haldwani (Naini Tal)	4.0	79.7
Madhya Pradesh	(1) Gwalior	4.4	79.0
	(2) Pachmarhi	4.0	77.8
Karnataka	(1) Kumta (North Kanara)	3.0	69.5

Propagation

The plant is sterile and so vegetative propagation by "rooted suckers" is preferred, which results in true to type plants. Stolons are also used for propagation. Planting method is similar to that of sugarcane, at 30–35 cm row space and planting 4–5 cm deep. Rooted suckers are laid end-to-end and covered with soil. 100–150 kg suckers are required to stock 0.405 ha. A hectare of well maintained *M. arvensis* field will yield 2470 kg of rooted suckers; suckers start sprouting within a week after planting and in 2 weeks sprouting is complete.

Planting Time

In north India mint plant is dormant in December–February period. It is best to plant during January–February.

As soon as the weather warms up, the suckers sprout. Planting sprouted suckers usually results in an erratic crop stand in the field.

Manuring

Best results are obtained through green manuring, using, *senji* (*Melilotus parviflora*), *dhaincha* (*Sesbania cannabina* syn. *S.aculeata*), alternatively, a basal application of 28 tonnes farm yard manure (fym)/ha is applied. The following Table 4 summarizes fertilizer use results.

Table 4 Effect of N, P and K application on yield of green herb of *M. arvensis*

Nutrient applied	Rate (kg/ha)	Herb yield (kg/ha)
N	Control	35.20
	50	40.39
	100	43.10
P	Control	36.70
	75	41.46
	150	42.84
K	Control	38.52
	75	39.84
	150	40.33

It is best to apply chemical fertilizers in three doses, first 40 days after transplanting, second and third application a week or 10 days after harvest of the herb foliage.

Interculture and Weeding

Thorough weeding is highly recommended. Manual weeding using a *khurpa* (a local hand implement like a sickle) is quite effective. The common *desi* (local) plough with a narrow bottom can also be used to weed in larger areas.

Among the herbicides, Diuron (0.6–0.8 kg/ha), pendimethalin (1–1.5 kg/ha) can be effective.

Harvest

Number of harvests depend, to some extent, on planting time. Investigations at the Regional Research Laboratory, in Jammu, in India, indicate the first harvest should be taken within 95 days after sprouting, which roughly corresponds to mid May. The gap between the first and second harvest should be reduced to 70–80 days, based on crop growth, and, the third harvest should be done towards the end of October or early in November.

Harvest in mint is a very delicate operation and negligence can lead to deterioration in quality of oil. Harvest should preferably be done on sunny days, and best time to harvest is when lower leaves turn yellow and start shedding.

Yield

Fresh yield averages about 14,800 kg/ha. The proportion of stem to leaves is 1:1, on average. The following Table 5 gives an estimate of herbage and oil yield.

Table 5 Herbage and oil yield

Locality	Oil percentage			Yield (kg/ha)
	Whole green herb	Whole dry herb	Dry leaves	
Jammu	0.45	1.36	3.42	30
Japan		1.48	3.78	27

With improvement in cultural and manurial practices, yields can be increased to as much as 99 kg/ha.

Crop Rotation

It has been found that the herbage yield (leaves/tops) decline in the third year after planting and it is economical to plant afresh. As a good measure of soil fertility conservation crop rotation is advisable and the following ones are recommended:

1. First and second year mint, summer fallow, wheat or pea in winter (*rabi* season) followed by mint
2. First and second year mint/sorghum (fodder crop) in summer (*kharif*) with berseem in winter (*rabi*) followed by mint
3. First and second year mint, rice followed by berseem and mint again
4. First and second year mint, rice with *dhaincha* (in low lying areas), followed by mint

Plant Protection

Rust caused by *Puccinia menthae* Pers., powdery mildew caused by *Erysiphe cichoracearum* DC, leaf blight caused by *Alternaria* sp. and wilt caused by *Verticillium albo-atrum* Reinke & Berth are the most prevalent and distinctive diseases of Mentha sp. They are briefly discussed below.

Rust observed all over the world where mint grows leads to defoliation and reduces yield by about 25%. Satisfactory control measures are yet to be in place, but, fungicidal sprays, treatment of suckers or runners by dipping them in hot water (45–46 °C) for 10 min before planting them can ensure some control, though, evolving resistant varieties is the best solution.

Powdery mildew appears early in winter and persists without much dissemination. Dusting finely ground sulfur or spraying wettable sulfur can be effective in disease control.

Leaf blight is fairly common in mint plantations. Spraying copper fungicides like Bordeaux mixture offers a fairly good control.

Wilt appears in summer. It causes leaf etiolation, dwarfing and ultimate plant death. It is a severe disease in the US, but, not so in India. No satisfactory control measures exist, but, deep ploughing before transplanting and using wilt-free planting material can ensure some control.

Minor Diseases

Anthracnose caused by *Sphaceloma menthae* Jenkins, leaf blight caused by *Cephalasporium* sp. and leaf spot caused by *Septoria menthae* Thum. are common overseas, not in India, though Fungi, such as, *Ramularia methicola* Sacc., *Rhizoctonia solani* Kuhn. *Helicobasidium purpureum* Pat. Syn. *Rhizoctonia violacea* Tul., *Phyllosticta menthae, Sclerotium rolfsii.* Sacc. and *Peranospora* sp. also cause many diseases in mint. They are of low importance.

Insect Pests

Our current knowledge of insect pests of mint plant and their control is inadequate. The following pests have so far been detected in India and the control measures are also discussed briefly, as follows.

Moths (*Syngamia abruptalis* **Wlk. Lepidoptera, Pyralidae)** The moths are orange colored with a wing span of 1.5 cm. Both the wings are traversed by chocolate-colored or brown undulating lines. The larva is about 2.5 cm in length and dull green in early stages, but, develops a yellowish color later. All along the body, on lateral sides, there are black spots. The larva folds a leaf length-wise along the midrib and webs the margins by strands, sometimes employing 2 or 3 leaves. It occupies this hide-out and scrapes the parenchyma, resulting in withering of leaves and ultimately the leaves fall off. Pupation takes place in the folded leaf in a white cocoon.

The insect attack is severe from July to October. The insect has a number of other hosts, such as, *Ocimum sanctum* and *Orthosiphon grandiflora* in South India; the pest is reported on *Mentha viridis* Linn.

No control measure has been perfected for this pest, yet.

Laphygma exigua **Hub; Lepidoptera, Noctuidae** The larvae of this pest feeds on leaf buds, by boring into them, thus making the leaves ragged and distorted with many holes. The caterpillars feed on younger leaves as they grow, and normally 4 or 5 larvae are found at the apex of the single branch. The pest attack is severe during spring, when the young leaves emerge. Among agricultural crops, the pest has a number of hosts, including onion, finger millet, lucerne and sorghum.

Autograpa nigrisigna **Wlk; Lepidoptera, Noctuidae** The caterpillars are more than 2.5 cm long. They feed voraciously during day and defoliate the plant. If defoliation is heavy, the plants often wither.

The attack is severe during March–April, after which the plants recover. This insect has also been observed feeding on other members of the Labiatae family. BHC 5% dust at 29.07–10.43 kg/acre is usually recommended as a control measure.

Euxoa segetums **Schiff; Lepidoptera, Noctuidae** Larvae of this pest feed at night on foliage and during day remain dormant in the loose clods of soil. They sometimes cut the growing shoots at ground level. Loss is heavy to the crop. Other host is potato crop. Chlordane 5% dust around plant base is recommended to control the insect.

Bihar Hairy Caterpillar (*Diactrisia obliqua* Walk; Accitidae, Lepidoptera) The pest has 2 or 3 generations/year. Larvae 3.75 cm long, clothed with tufts of long hair in early stages. The pest attack is usually severe during rainy season. DDT 50% wettable powder @0.45 kg in 68.18 liters of water or 1% parathion dust @6.803 kg/0.4 ha is recommended as a control measure.

Bug (*Monanthia globulifera* Walk; Tingidae; Homoptera) Popularly known as the "Lacewing bug", the pest severely damages the mint crop and also *Ocimum sanctum*. The bug causes foliage to curve, wilt and finally die. An outbreak of this pest can be quickly recognized by blackish-brown spots on the leaves. The severity of attack mostly depends on the favorable environment. Dusting of BHC 5% protects the crop if the crop is grown on a small scale.

Economics of Mint Oil Production The production cost of mint oil in India ranges between INR 92 and 123/kg which is considerably low compared to that in other mint-growing countries, because of cheap labor and low input costs in India. Technological improvements, in terms of improved varieties and management practices will reduce cost of cultivation. The following Table 6 briefly summarizes some of these aspects.

Strategies to Minimize Cost of Mint Oil Production

1. Use of improved strains/variety:

The cost of production per unit of mint oil is substantially lowered in case of improved strains, as compared to the local ones. This is mainly because of the higher oil yield achieved through plant breeding. The results in the above Table 6 substantiate this. The higher oil yield of MSS 1 is mainly due to considerable

Table 6 Economics of local and improved spearmint strains

Details	Local strain	Improved strain (MSS 1)	Percentage increase
Herb yield (tonnes/ha)	1.35	1.90	40.7
Oil content (%)	0.29	0.56	93.1
Oil yield (kg/ha)	36	98	272.2
Production cost of oil (INR/kg)	180	77	47.78 less

Source: Singh et al. (1984)

improvement in biomass production and oil concentration in green herb. Likewise, remarkable improvement was obtained in Japanese mint (MAS 1, Hybrid 77), Bergamot mint (Kiran), and, Scotch spearmint (MCAS 2). The planting materials of all improved strains are now with the Indian farmers.

2. Optimum planting time:

Optimum time of planting depends on the prevalent climate. Maximum herbage and oil yield in Japanese mint were obtained with mid January to early February planting under western Uttarakhand conditions (Singh et al. 1986). For other mint species it is early January (Prasad and Saxena 1980). In the temperate climate of Himachal Pradesh, March was found to be the best time for planting spearmint (Katoch et al. 1979). Under delayed planting, when mint is planted after mustard/wheat harvest in March–April, transplanting of seedlings raised in nursery beds is highly beneficial to minimize yield reductions.

3. Application of optimum quantities of fertilizers and irrigation: Mint responds well to fertilizers, in particular, to N fertilizers. Maximum herb and oil yield were obtained when the crop was irrigated at 10 days interval during summer months comprising the period of first harvest (Duhan et al. 1977). During the period after first harvest, irrigation requirement depends on rainfall. The following Table 7 shows the fertilizer (N)-irrigation interaction on mint yield and net return

The above table clearly demonstrates that high mint yield is obtained when application of N is accompanied by adequate irrigation. In Japanese mint, maximum response per kg N, net return and net return per INR invested were obtained with 50% available soil moisture level (ASM) and 172 kg N/ha. Mint cannot withstand waterlogging and therefore, good drainage is a prerequisite for higher oil production. Similar to N application, P application also increases yield of Japanese mint (Kothari et al. 1987a). The optimum dose of P varies with location and was found to be 35 kg/ha in the *Tarai* region of Uttarakhand (Kothari et al. 1987a). Mint plant, however, does not respond to K under Indian conditions (Rao et al. 1983), because, Indian soils are invariably rich in K content.

Table 7 Effect of irrigation on N requirement and economics of Japanese mint

Irrigation level	Optimum dose of N (kg/ha)	Yield at optimum N (kg/ha)	Response at optimum N (kg/ha)	Response/ kg N	Net return at optimum N (INR/ha)	Net return per INR investment
No irrigation	100	97	64	0.64	3440	0.31
25%	124	173	100	0.81	11, 460	0.79
ASM[a]						
(8)[b]						
50% ASM (14)	172	221	144	0.84	15,870	0.92
75% ASM (19)	168	223	140	0.83	15, 750	0.89

Note: [a]*ASM* Available Soil Moisture; [b]Number of irrigations required to maintain desired moisture level in the field

Weeding and Mulching

Weeds are of major concern in mint farming and reduction in oil yield due to weeds ranges from 60% to 80%. In general, weeds are removed manually from mint fields, and, for obtaining two good harvests about 4–5 weedings are required (Gulati and Duhan 1971). Mint field should remain relatively weed-free during critical periods of growth, namely, 30–75 days after planting and 15–45 days after first harvest (Kothari and Singh 1991). The following Table 8 summarizes some of the results with weedicides.

Adoption of Suitable Cropping Systems

Mints are grown in areas with assured irrigation facilities, and therefore, multiple cropping is a common practice. Sequential cropping of maize-potato-mint or maize-*lahi*-mint was found to be highly remunerative (Singh and Singh 1978). In areas of high rainfall or the *Tarai* region of Uttarakhand, paddy-mint sequence has been recommended (Dutta 1971). Mint is also grown in large areas of the *Tarai* region of Uttarakhand where sugarcane is also extensively grown. Since initial growth of the sugarcane crop is slow, mint can be grown as an intercrop (Kothari et al. 1987b). It is clear that intercropping may bring down substantially cost of mint oil production. There is also scope to grow mint in plantation of tree crops like poplar, subabul (Singh et al. 1990).

Harvest at Optimum Time

Time of harvest, in general, has a close relationship with yield and quality of the mint oil. The cycle of leaves, their origin, expansion of leaf development and finally loss through senescence, are particularly important in mint leaf as leaf contains

Table 8 Economics of manual and herbicidal weed control in Japanese mint

Treatments	Oil yield (kg/ha)	Gain over weedy check (kg oil)	Cost of weed control treatments (INR/ha)	Net return per INR investment
Oxyfluorfen 0.25 kg				
a.i/ha	154	88	1000	12.2
Pendimethalin 0.75 kg				
a.i/ha	149	83	900	12.8
Hand-weeding (Twice)	150	84	1200	9.5
Weed-free (4 hand-weedings)	159	93	2000	6.0 6.0
Weedy check	66			

Modified from Kothari and Singh (1993)

agronomically valuable product. In general, first harvest of mint is done 110–120 days after planting; second harvest 60–70 days after the first. It may vary considerably depending on soil, climatic conditions and the mint species cultivated. In spearmint, harvest at flowering gave maximum oil yield under north Indian conditions. Japanese mint and Bergamot mint do not bloom, and therefore, it is difficult to precisely determine the proper harvest stage. Several empirical rule of thumb are, therefore, in practice, like, cessation of growth, ratio between leaf shed at the bottom and new ones emerging at the top, in which case harvest is done when the value exceeds 1. Mint crop should be harvested in bright sunny weather (Kothari and Singh 1994).

Post Harvest Technology

The crop can be dried on the field itself by spreading the plants evenly on the field bunds, provided the weather is sunny. The plants should not be allowed to dry completely; if so, leaves drop off while stacking the plants or during transport. It is advisable to collect the herbage within 12 h of spreading. After harvest, the plants should, preferably, be tied into bundles, hung in the open air or in closed sheds and dried.

Processing Technology

In all mints and spearmints, the volatile oil occurs in numerous minute glands on the upper surface of the leaves.

Stem contains only traces of oil. The percentage of essential oil on dry weight/volume basis shows a clear-cut oil gradient in different plant parts. A partitioning of this is as follows:

Unexpanded leaves: 9.25%
Expanding leaves: 9.31
Fully expanded leaves: 4.04%
Yellowing leaves: 4.04%
Flower buds: 7.52%
Flowers: 6.73%
Seed-bearing portions of the inflorescence: 5.17%
In stem, the partitioning is as follows:
Upper tender portion: 0.51%
Middle portion: 0.21%
Lower woody portion: 0.08%

It may be concluded from the above-mentioned details that younger and tender plant parts contain the largest amounts of oil. With the advancement of leaf age and increase in size oil does not increase proportionally because the number of glandular hairs remain the same, except that they are filled fully in mature leaves.

Observed that application of GA (Gibberlic Acid) at the pre-flowering stage induces and increases total herbage yield from 35% to 50%, but, the essential oil tends to decrease by 5–15%. It has no effect on menthol percentage.

Distillation of dried leaves works out cheaper than distilling fresh leaves. By steam distillation and filtration a golden-yellow-colored volatile oil is obtained. Leaves and flowering tops give the highest amount. About 50% menthol can be separated out in crystalline form on cooling the oil. The remaining (dementholized) oil is used as peppermint oil. This process has a great commercial potential.

Peppermint oil is stored in colored bottles, in airtight aluminum or galvanized containers in a cool dry place. Presence of moisture in the oil can rancidify it. Since acidic, it should not be stored in tin containers.

Essential Oil Distillation

The sizes and types of modern stills differ according to the quantity of material to be distilled, and type of heating system to be adopted – direct or steam. Mint is a crop, which, for economy in transport and handling should be distilled at the site of the farm. The field still should be designed and operated according to the size of the mint farm. In case of small individual holdings, it is advisable to have a still on a co-operative basis. The Regional Research Laboratory, (RRL), Jammu, India, have designed some types of distillation stills, which are found to be quite efficient.

Composition of Oil

The natural oil yields, on an average, 40–50% of menthol and 50–60% of dementholized oil which can be used both in confectionery and medicine in place of imported peppermint oil. Japanese mint oil is not distinguished from peppermint oil in the Indian trade. The dementholized oil has been found to contain methyl acetate (24.4%), free menthol (44.6%), menthone (24.6%) and hydrocarbons (6.2%). Among the hydrocarbons, alpha-pinene, alpha -1limonene, caryophyllene and cademens are present.

The quality of the Japanese mint oil grown in Jammu and Uttarakhand, in India, compares favorably with that obtained from Japan and Brazil. It's physico-chemical properties are as follows:

Specific gravity at 25 °C: 0.8969–0.9903
Refractive index at 25 °C: 1.4494–1.4573
Optical rotation: 41.0^0 2′
Congealing point: 15.0

Acid value: 1.5–2.8
Ester value: 14.12–29.47%
Total menthol: 61.3–94.4%
Solubility in 1.5–2.0 volumes of 70% alcohol

Economics of Distillation

Experiments carried out at Regional Research Laboratory (RRL) in Jammu, in India, have indicated that for a known volume of oil distilled, fuel consumption, duration and labor employed in the case of fresh herb are much higher in comparison to air-dried herb. It was further observed that it is economical to distill dried leaves alone, as stems contain only very little (traces) of oil (Table 9).

Constraints in Essential Oil Production

The oil industry requires the following support:

1. Involvement of Agriculture and Horticulture Departments of the Government of India to promote aroma crops like mint crops
2. Development of high-yielding strains by R&D institutes
3. Interaction between growers, scientists, and processors/manufacturers and end users by establishing a high-level committee involving associations, R&D institutes and government agencies
4. Marketing support and overseas publicity by CHEMEXCIL
5. Support price to famers by Government Agencies wherever supply exceeds demand by establishing an Essential Oil Board or Society

The recent discussions with farmers of Northern Indian States have confirmed that present prices of many essential oils are not sufficiently conducive for enhanced production. For instance market for sunflower oil is assured and much better compared to mentha oil.

Table 9 Progress in essential oil production in India *vis-à-vis* world production

	Production in India (tonnes)			
Essential oil	1982	1992	World production (tonnes)	Major producing countries
Mint oil	400	5000	13,000	India, China
Peppermint oil	10	100	2000	USA
Spearmint oil	2	150	1500	USA, China
Mentha citrata oil	10	80	200	India, USA

Uses

Mint or *pudina* (in Hindi language) is very popular for use in the common *pudina-dhania* (coriander)- *chutney*. Mint is also used to flavor meat, fish, sauces, soups, vinegar, teas, tobacco, and cordials. The fresh leaf tops of all the mints are used in beverages, fruit cups, apple, sauce, ice-cream, jellies, salad, sauces for fish and meats, also to flavor vegetables. Roast lamb and mint jelly have become inseparable companions. Japanese mint is also used as a substitute for true peppermint oil (from *M. piperita*), which resembles it in physico-chemical properties. It possesses a somewhat bitter flavor and is considered inferior to *M. piperita* in aroma and quality. Arvensis oil with low menthol content is finding some use in cheap perfumery.

Menthol Production

Japanese mint oil finds uses similar to that of peppermint oil. However, the peppermint oil is preferred for flavoring purposes. It is used for the production of natural menthol. Dementholized oil is employed to flavor mouthwashes, toothpastes and pharmaceutical preparations. The main use of mint is the extraction of volatile oil which contains menthol. Singh and Sharma (1994) have provided a ready reckoner for the quick estimation of menthol concentration in mint oil.

In Medicine

It is used to cure stomach disorders, in ointments for headaches, rheumatism and other pains, cough syrups, inhalations, mouth washes, toothpastes etc. The harsh flavor of *arvensis* is masked to some extent by skillful blending of the two oils. Of course, *arvensis* is not used where delicacy of aroma and flavor are the prime objectives.

The oil and the dried plants are antiseptic, carminative, refrigerant, stimulant, and diuretic. The dried plant does not have a good taste, as it is, but, is expectorant, emmenagogue, tonic to the kidney, useful in curing the diseases of the liver and spleen, asthma etc. It also possesses anti-spasmodic properties, is used in jaundice, and frequently given to stop vomiting.

In China, the leaves and stems are made into an infusion and used as a carminative, sudorific and anti-spasmodic. In Assam State, India, the plant is considered as an excellent diaphoretic. An infusion is given in fever, indigestion etc.

Synthetic menthol is also now being manufactured in a number of countries, but, its flavor is not equal to the flavor of the natural one.

References

Dutta PK (1971) Cultivation of Mentha arvensis in India. Flavour Ind 2:233–240

Duhan SPS, Singh VP, Bhattacharya AK, Hussain A (1977) Response of Japanese mint to different irrigation schedules. Perfum Flavor 2(5):57

Gulati BC, Duhan SPS (1971) Japanese mint in Nainital *tarai* of Uttar Pradesh. Indian Perfum 15:1–14

Katoch PC, Bharadwaj SD, Kausal AN (1979) Effect of time of planting and row spacing on herb yield and oil content of spearmint. Indian Perfum 23:91–94

Kothari SK, Singh K (1991) Critical periods of weed interference in Japanese mint (*Mentha arvensis* L.). Trop Pest Manag 37(1):85–90

Kothari SK, Singh K (1993) Chemical weed control in Japanese mint. (*Mentha arvensis* L.). J Essent Oil Res 84

Kothari SK, Singh UB (1994) Economics of mint oil production. Indian Perfum 38(1):15–22

Kothari SK, Singh V, Singh K (1987a) Effect of rates and methods of phosphorus application on herb and oil yields and nutrient concentrations in Japanese mint (*Mentha arvensis* L.). J Agric Sci (Camb) 108:691–693

Kothari SK, Singh V, Singh K (1987b) Intercropping of mint species in spring planted sugarcane under *tarai* conditions of Uttar Pradesh. Indian J Sugar Technol 4:1–6

Prasad S, Saxena MC (1980) Effect of dates of planting and row spacing on the growth and development of peppermint in *tarai*. Indian J Plant Physiol 23:119–126

Rao BRR, Rao EVSP, Singh SP (1983) Influence of N, P and K fertilization on the herbage yield, essential oil content and essential oil yield of bergamot mint (*Mentha citrata* Ehrh.). Indian Perfum 28:105–107

Singh AK, Sharma S (1994) Quick estimation of menthol concentration in the essential oil of Mentha arvensis. A ready reckoner. Indian Perfum 38(3):93–97

Singh JP, Singh JN (1978) Nitrate and ammonium as sources of nitrogen for Japanese mint and their influence on the uptake of other ions. Indian J Agric Sci 48(5):274–278

Singh A, Singh DV, Husain A (1984) A promising strain of spearmint (*Mentha spicata* L.). Indian Perfum 28(3–4):142–145

Singh K, Singh V, Kothari SK (1986) Effect of planting material and spacing on herb, oil and sucker production in *Mentha arvensis* L. Ann Agric Res 7:313–316

Singh VP, Kothari SK, Singh K, Singh DV (1989) Effect of irrigation and nitrogen on herbage and oil yield of Japanese mint (*Mentha arvensis* L.). J Agric Sci (Camb) 113:227–229

Singh K, Singh V, Husain A, Kothari SK (1990) Aromatic plants as efficient intercrops under poplar (*Populus deltoides* Bartram Ex. Marshall). Indian Forester 116:189–193

Spices Board, Govt. of India (1988–96) Annual statement of exports of spice and spice production from India

Peppermint

Abstract The chapter will begin with a description and distribution of the plant. Subsequently, area, production, exports and crop management aspects would be discussed. There would be an elaborate discussion on pharmaceutical uses of mint oil.

Keywords Description · Distribution · Pharmaceuticals · Waste utilization

Botanical name: *Mentha piperita* Linn.

Description and Distribution

Mentha piperita is the potential hard currency earner for a country like India. Of late, it's demand has increased manyfold, consequently, the price of peppermint has crossed over INR 1000/kg.

M. piperita is extensively cultivated in USA, Australia, England and in European countries. The plant was cultivated in the Nilgiris, in Tamil Nadu, in 1818 and since then, it has been raised in Karnataka, Jammu and Kashmir, Delhi and Dehra Dun. Peppermint is a perennial, glabrous, strongly scented herb grown or cultivated in temperate regions of Europe, Asia etc. It is considered to be the hybrid between *M. spicata* and *M. aquatica*. It is sensitive to drought. It is erect, 30–90 cm high, purplish or green, leaves are ovate or oblong, coarsely serrate, smooth and dark green above, pale or sparingly hairy below; flowers are purplish in thick terminal spikes. The herb consists of dried leaves or crushed dried leaves of the cultivated peppermint plant.

K. P. Nair, *Herbal and Acidulant Tree Spices*,
https://doi.org/10.1007/978-3-031-04360-4_12

Area, Production, and Export/Import Markets

India does not produce enough peppermint for the manufacture of peppermint oil to meet the country's internal demand. Thus, during 1981–82, of the total import of spice essential oils worth INR 10.9 million, the import of peppermint oil alone amounted to INR 8.6 million, which amounts to 79.9% of the total expenditure.

Crop Management

Method of cultivation is almost the same as described in the case of Japanese mint. Leela and Angadi (1994) observed that essential oil content and menthol are the highest and methane content was minimum from July harvest. It is suggested to harvest the herbage in July to obtain high yields of good quality oil.

Processed Products

Volatile Oil

The herb is the source of true peppermint oil, extensively used to flavor and also in manufacture of pharmaceutical products. Commercial oils are derived from cured, partially dried plants, while official oils are obtained from fresh material; the oil is rectified, if necessary. The yield of oil varies from 0.3% to 1% depending on the extent to which the material has been dried before distillation. Dried leaves and flowering tops from Kashmir gave 0.7–1.0% oil, while fresh mature leaves from Kanpur gave 0.32% oil.

Physico-chemical Properties

Peppermint oil is a colorless, pale-yellow or greenish-yellow liquid with a strong agreeable odor and a powerful aromatic taste, followed by a cooling sensation when air is drawn into the mouth. On ageing, the oil darkens in color and becomes viscous. When chilled, menthol separates as crystals. The oil contains menthol (50–55%), methyl acetate, menthone and small amount of several other compounds. The peppermint oil steam-distilled from Kashmir-grown plants had the following properties:

Specific gravity at 15 °C: 0.9187
Refractive Index; 1.4666
Methyl acetate: 14.4%
Soluble in 6 volumes of 70% alcohol

Quality Standard

The Indian (BIS) physico-chemical specifications for peppermint oil are set out in the following Table 1.

Uses

Medicine/Pharmaceuticals

The herb is considered aromatic, stimulant, stomachic and carminative and is used for allaying nausea, flatulence, and vomiting. Bruised leaves are used as an external application to relieve local pains and headache. A hot infusion is taken to cure stomach ache and colicky diarrhoea. The drug is frequently adulterated with spearmint, which is difficult to detect.

Peppermint oil is one of the most widely used essential oils. It is used to flavor pharmaceuticals, dental preparations, mouth washes, cough drops, confectionery and alcoholic liquors. It is used internally and externally and is preferred to menthol

Table 1 Physico-chemical properties of peppermint oil

Characteristics	Requirements		Test method
	Type 1	Type 2	
Color and appearance	Colorless, pale yellow or greenish yellow	Pale Yellow	IS:326 (Part 2) 1980[a]
[b]Odor and taste	Characteristic strong minty, herbal, followed by cooling sensation	Characteristic minty, intense herbal, followed by cooling sensation	IS:2284 1988[a]
[c]Rel.density at 27/27 °C	0.8773-0.9123	0.8923–0.9123	IS:326 (Pt-3) 1980
Optical rotation	−35 to −45°	−20 to −40	IS:326 (Pt-4) 1980[a]
[d]Refractive index [d]27 °C	1.4562–1.4642	1.4512–1.4632	IS:326 (Pt-5) 1986[a]
Free alcohols as 1-menthol	Minimum 60	20	
(mol.wt.156.26)%, m/m per cent by mass	Maximum	30	
Esters, as methyl acetate (mol.wt. 198.28)	3-15	5-30	IS:326 (Pt-8) 1980[a]
Total alcohols	65 Min	5 Min	IS:326 (Pt-9) 1980[a]
Ketones, as methone (mol. wt. 154.25)% m/m (using free hydroxylamine alcohol)	5–20	20–40	IS:326 (Pt-II) 1986[a]

Source: BIS/PCDC 18 (68.6)-1988

[a]Methods of sampling and test for natural and synthetic perfumery materials

[b]Methods for olfactory assessment of natural and synthetic perfumery materials

[c]The correction factor for each degree Celsius change in temperature is (first.rev.) 0.000038

[d]The correlation factor for each degree Celsius change in temperature is 0.00064

because of its more pleasant taste. The oil has mild antiseptic and local anaesthetic properties. Externally, it is used against rheumatism, neuralgia, congestive head-ache and tooth-ache.

Waste Utilization

The green plant, after extraction of the essential oil may be dried into hay or silage for use as cattle-feed. The hay contains 12.7% protein, 8.5% digestible protein, 49.4% total digestible nutrients, and, has a nutritive ratio of 4.8%. It may be employed as a substitute for lucerne-hay to feed dairy cows.

Reference

Leela NK, Angadi SP (1994) Yield and quality of peppermint oil as influenced by time of harvest. Indian Perfum 38(4):120–122

Spearmint

Abstract As usual, the chapter will begin with a discussion on description and distribution of the plant. It would be followed by area, production and exports/imports, followed by crop management techniques. There would be an elaborate discussion on processed products, especially composition of volatile oil and its uses.

Keywords Description · Distribution · Processed products · Volatile oil · Composition of oil

Botanical name: *Mentha spicata* Linn.
Syn. *Mentha viridis* Linn.
Family: Labiate

Description and Distribution

Spearmint belongs to the mint family and is a relative of mint and peppermint, described earlier. It is a glabrous perennial, 30–90 cm high, with creeping rhizomes, indigenous to the north of England, but, grown all over the world. It is cultivated in Indian gardens, in the plains. Leaves are smooth or nearly so, are sessile, lanceolate to ovate, acute and coarsely dentate, and are smooth above and glandular below. Flowers are lilac, in loose, cylindrical, slender, interrupted spikes. The leaves have a characteristic aromatic odor and slightly pungent in taste and when eaten do not leave a cooling sensation, as in the case of peppermint (please see Table 1, above in chapter "Peppermint").

Originally, spearmint was a native of Central and Southern Europe, but, the present day commercial cultivation of the plant and distillation of oil are carried out in the far western USA, especially in the State of Oregon and Washington. Introduction of oil of *Mentha citrata* into India is largely attributed to the efforts of the Regional Research Laboratory (RRL) in Jammu, India, in the early seventies.

K. P. Nair, *Herbal and Acidulant Tree Spices*,
https://doi.org/10.1007/978-3-031-04360-4_13

The commercial scale cultivation of *Mentha citrata* in India is of recent origin, centered chiefly in Jammu and Uttarakhand States especially in Kashipur, Budaun, Moradabad and Haldwani in the latter. The essential oil of *Mentha citrata* is produced by steam distillation of fresh green crop which is partially dried in shade. The average yield of oil has been observed to be around 40 kg/acre.

The current annual indigenous production of the oil of *Mentha citrata* is estimated to be in excess of 5000 kg in India. The oil is mainly consumed within the country for perfumery purposes.

Scope In view of its potential in the perfumery industry, there is considerable scope for increasing production of this oil in India. Efforts are being made to extend cultivation of *Mentha citrata* to suitable centers, in particular, in the States of Punjab, Haryana and Karnataka (in Bengaluru). If the attempts are fruitful, India would easily increase the production of the oil three to four-fold, in the coming decade. Hence, if good quality oil is produced internally, in India, sufficient surplus would be there for export after meeting the domestic demand.

Crop Management

Cultivation method is almost similar to that of Japanese mint, described earlier. It is planted in the month of January. Two or three harvests are obtained in the months of May-June, July-August and October–November.

Important steps in the successful cultivation of Spearmint in India:

If the following steps are taken in the cultivation of Spearmint in North India, its good quality oil, acceptable to the world trade, can be produced with good economic returns.

1. The planting must be completed by the end of the second week of January
2. Take the first cutting during the later half of May to early June. During this phase, keep a close watch on leaf shedding while determining the time of harvest so that the economic yields of the fresh herb are obtained
3. Avoid over-irrigating the crop, especially from the end of April until the harvest of the crop
4. Stop irrigating one week before harvest
5. Distil the crop in the fresh state for better quality of oil and higher returns

The spearmint fresh leaf has the following composition:

Moisture: 83%
Protein: 4.8%
Fat: 0.6%
Carbohydrates: 8%
Fiber: 2%
Mineral matter: 1.6%

Calcium: 200 mg
Phosphorus: 80 mg
Iron: 15.6%

Vitamins

Carotene (as Vitamin A): 2700 IU (International Units)
Nicotinic acid: 0.4 mg
Riboflavin: 80 mg
Thiamine: 50 mg/100g
The leaves contain traces of copper (1.8 mg/g).

Processed Products

Volatile Oil

The fresh flowering herb, on steam distillation, yields 0.25–0.50% of volatile oil, known as spearmint oil. It is a colorless, yellow or greenish - yellow liquid with the characteristic odor and taste of spearmint; the aroma improves on ageing. The characteristics of spearmint oils obtained on an experimental scale at different places in India are as follows:

The main characteristic constituent of the oil is 1-carvone. Oil distilled in Kanpur has the following properties:

Carvone: 55.8%
Terpenes: 17.5% (chiefly α-limonene and dipentene):

Alcohol (as dihydrocarveol): 6.7%
Esters: 11.6% (as dihydrocarveol acetate)

A sample of oil distilled from plants grown in Pune contained no carvone at all. These differences in composition of spearmint oil are evident due to varietal or specific regional variations.

The essential oils obtained from the Indian spearmint (*Mentha spicata*) subjected to different drying procedures were analyzed employing gas chromatography, and mass spectrometry techniques. Forty three flavor components were identified, of which, eucarvone, cubenol, and α-cadinol have been reported for the first time. The essential oil had a higher percentage of limonene (26.83%), compared to Italian and American oils. Carvone to Limonene ratios were 2.2:1.0 and 2.3:1.0 in the oil from fresh and shade-dried spearmint, respectively. Blanching leaves prior to drying yielded products which were unattractive with respect to color and appearance and were also bland and odorless due to loss of volatile oil during blanching and drying.

Storing fresh leaves for 12 h after spraying with water increased volatile oil content by about 6%. Shade drying of leaves resulted in a product with good green color and minimum loss of volatile oil compared to other drying methods.

Composition of Oil

Specific gravity at 15 °C: 0.925–0.940
Optical rotation: −30 to −50^0
Refractive Index: 1.4800–1.4900
Carvone content: 42–60%

The oil has linalyl acetate and linalool as the main constituents and is employed in the preparation of perfumes with bergamot/lavender properties.

Oil of Mentha citrata is soluble in 1.5–2 volumes of 70% ethyl alcohol (v/v), occasionally with slight opalescence. The BIS specifications for this oil are given in the following table (Table 1).

Packing

The material shall be supplied in air tight, preferably, amber-colored glass, aluminum or tin-plate containers, permitting a minimum of air space, or as agreed to between the supplier and the customer.

Table 1 Physico-chemical specifications/requirements for oil of *Mentha citrata*

Physico-chemical characteristics	Requirements
Color and appearance	Colorless to pale yellow
Odor	Sweet-herbaceous with fruity/citrus notes (resembling lavender/bergamot with clary sage undertones)
[a]Relative density at 27/27 °C	0.8797–0.8897
Optical rotation	−7 to −8^0
[a]Refractive Index at 27 °C	1.4559–1.4609
Acid value, max. (using 10 g of sample)	1.0
Esters calculated as linalyl acetate ($C_{12}H_{20}O_2$ 25%) by mass, min	
Total alcohols calculated as linalool ($C_{10}H_{18}O_2$ 80–90%) by mass	
Linalyl acetate, per cent by mass, min 25	
Linalool, per cent by mass, min 40	

Source: BIS-PCDC 18((25)P-1988
[a]The correlation factors for relative density and refractive index for each degree Celsius change in temperature are 0.00064 and 0.00038 respectively

Storage

The packed spearmint oil should be protected from incident light and stored in a cool and dry place.

Uses

Green leaves of the plants are used for making *chutney*, and to flavor culinary preparations, vinegar, jellies and iced drinks.

The herb is considered a stimulant, a carminative and an antispasmodic. A soothing tea is brewed from the leaves and an alcoholic beverage (*mint julep*) is prepared from them and used as an antidote against poison. A sweetened infusion of the herb is given as a remedy for infantile problems, vomiting in pregnancy and hysteria. The leaves are used in fevers and bronchitis. Spearmint oil is used to flavor chewing gums, tooth-pastes, confectionery and several pharmaceutical preparations.

Origanum

Abstract As usual, the chapter will begin with a description and distribution of the plant. It would then be followed by discussion on area, production and trade, post harvest technology and end uses.

Keywords Description · Distribution · Post harvest technology · Food flavorant · End uses

Botanical name: *Origanum vulgare* Linn.
Family: Labiatae

Description and Distribution

Origanum, oregano, Mexican origanum, the "pizza herb", "origany", or wild marjoram of trade is one of the most popular herbs in Mediterranean cooking. It constitutes the dried leaves of an aromatic, branched, perennial herb, 30–90 cm high, found in the temperate Himalayas from Kashmir to Sikkim, at an altitude of 1500–3600 m above mean sea level. The plant is used as a pot-herb or vegetable in Lahaul, Punjab, and Kashmir regions. Leaves are broadly ovate, entire or rarely toothed, about 1.5 cm long, and light green when dried; flowers are purple or pink in corymbose cymes; nutlets are smooth, brown. Oregano is cultivated in Greece, Dominican Republic, Turkey, Sicily and Italy. It grows abundantly in Mexico and is also known as "Mexican sage". It is a perennial herb of the mint family. It has a creeping root system and a rather sprawling habit. *Origanum vulgare* is very common in Shimla Hills and Kashmir Valley, in India.

The color of the dried herb is light green. The aroma of the herb is strong camphoraceous and resembles that of marjoram. The aroma varies according to the soil and climate. In fact, the Spanish word "oregano" means "marjoram". The taste is fragrant, spicy, warm, pungent ad bitter. According to some, the properties of Mexican origanum resemble those of sage rather than those of marjoram. The plant

K. P. Nair, *Herbal and Acidulant Tree Spices*,
https://doi.org/10.1007/978-3-031-04360-4_14

owes its usefulness as a culinary herb to its volatile oil. The herb is traded both as "whole" dried leaves and in "ground" form.

Area, Production and Trade

There is no official and reliable data on these aspects. However, there is considerable world trading of origanum or oregano notably in the Western countries. Though reliable figures on its world production is unavailable, Mexico alone exported the herb to the tune of 1349 tonnes valued at US $ 838,000 of which 90% was exported to the USA in as early as 1977 and Peru exported to the tune of 99.5 tonnes valued at US $ 735, 000. The USA also imported oregano oil from a number of countries to the tune of 2580 tonnes valued at US $ 3, 662,000. During the last five decades the trade has tripled. Hence, India stands to gain much in oregano production for the cash it would earn through the export of oregano oil.

Crop Management

Information on crop management of Origanum is rather scanty, but, the following is a brief narrative of the same.

Soil and Climate

It is a hardy plant and can be grown in all warm garden soils. Temperate to subtropical climate is needed. It grows in light, well-drained soil in a sunny climate.

Propagation

It is propagated by seeds, cutting, layers, or by root division. Seeds may be sown in spring or the roots may be divided and planted in autumn, winter or spring season.

Sowing

Seed can be sown during October in the plains and during March and April in the hills.

After-Care

The plant should be well nourished (through organic manures and/or chemical fertilizers) and looked after. Only then will it multiply to grand proportions and also yield well, generating good profits to the grower.

Harvest

The leaves and tops are cut prior to blooming, dried in shade and are used to flavor foods in the same manner as sweet marjoram (*Marjorana hortensis*). The plant is used in Punjab as a pot-herb. It is eaten also as a vegetable in Lahaul. The Origanum plants collected near Gulmarg (in Kashmir) on steam – distillation yield 0.76% of an essential oil.

Post Harvest Technology

Drying

Oregano should be dried when in flowering in order to allow full development of its aroma or flavor. Stalks may be picked as they blossom, and hung up in small bunches in shade for drying. Dried leaves are rubbed at the time of use to garnish foods.

The dried leaf is about 1.5 cm long and is light green in color. Generally, it is available as rubbed (or crushed) or in ground form. It is an important pizza spice. Its quality is measured by its volatile oil content, moisture, total and acid-insoluble ash.

The composition of Oregano is as follows:

Moisture: 80%
Protein: 11.7%
Fat: 6.4%
Crude Fiber: 11.0%
Carbohydrates: 53.9%
Total Ash: 9.0%
Acid-Insoluble Ash: 1.0%
Calcium: 1.7%
Phosphorus: 0.20%
Iron: 0.05%
Sodium: 0.02%
Potassium: 1.7%
Vitamin A: 1010 IU (International Unit) per 100 g
Vitamin B_1: 0.34

Vitamin B_2: 0.41
Niacin: 6.2
Vitamin C: 12 mg/100 g
It's calorific value is 360 calories/100 g.

Processed Products

Volatile Oil

Origanum contains volatile oil, fixed oil, cellulose, pigment, and mineral elements. Oil of Origanum appears to be sold under this name from a number of related species of Labiatae plants. In some Origanum oil, thymol is the main constituent, in others, it is carvacrol. Botanists and herb-growers classify it as "pizza herb".

The composition of the herb is as follows:

Volatile oil: 0.15–0.40%
Tannin: about 0.8%

The oil of European origin (specific gravity at 15 °C: 0.868–0.910; optical rotation: −20 °C to 70 °C) possesses an aromatic spicy, somewhat basil-like odor and contains thymol (up to 7%), carvacrol, free alcohols (about 13%), esters as geranyl acetate (2–3%) and bicyclic sesquiterpene (12.5%). Steam – distillation of the whole plant from Kashmir, India, gave a pale-yellow oil (yield 0.2%) with a pleasant smell and the following characteristics:

Specific gravity at 27 °C: 0.8812
Refractive index at 27 °C: 1.4795
Optical rotation: −1.5°
Acid value: 2.5
Ester value: 10.4 (after acetylation: 102.7)
Phenol content: Nil, freely soluble in 90% alcohol. It contains 1-pinene, dipentene, linalool, bi and tri-cyclic sesquiterpenes and palmitic acid. The oil called "Oil of Origanum" in trade, is really "Thyme Oil" (from *Thymus vulgare* Linn.)

Oil of *Origanum vulgare* is often confused with sweet marjoram oil (from *Marjorana hortensis*, which, however, is dextro-rotatory (up to +40).

Uses

As a Food Flavorant

Oregano is an essential ingredient of chilli powder, and is used, in "chilli concarne" and many other Mexican dishes. It is the spice that made pizza famous is equally good in any tomato-type dish, from spaghetti to old fashioned stewed tomatoes. Origanum is also used to flavor soups, meat dishes, pork, fish, egg and salads.

It is reported that no Mexican kitchen is complete without Oregano since they believe that no other herb imparts or communicates such an excellent aroma and flavor to food. The leaves and tops, prior to blooming, are used to flavor foods in the same manner as sweet marjoram (*Marjorana hortensis*). The plant is used in Punjab, India, as a pot-herb. It is eaten also as a vegetable in Lahaul. It was formerly employed to flavor ale and beer before hops were introduced in brewery industry.

In Medicine

The oil possesses carminative, stomachic, diuretic, diaphoretic, and, emmenagogue properties; it is given as a stimulant and tonic in colic diarrhoea; it is also applied in chronic rheumatism, tooth and ear ache. Due to the spasmolytic action of the oil, it is used in whooping cough and bronchitis. In homeopathy, it is used for hysteric condition. It is used as an external application in healing lotions for treating wounds, usually in conjunction with other herbs. The oil has been employed in veterinary liniments. It is used in gargle and bath. It stimulates growth of hair. The oil is used in the cosmetics and soap industries.

Other Uses

The herb is reported to be cultivated in eastern European countries like Poland for its seeds from which a fatty oil is extracted; seeds yield about 29.2% of the fatty oil.

Parsley

Abstract The chapter will at first discuss description and distribution of the plant. Then it will go on to discuss area, production, trade, crop management post harvest technology and discuss further the processed products and their uses.

Keywords Description · Distribution · Post harvest technology · End uses

Botanical name: *Petroselinum crispum* (P. Miller) Nymann ex. A.W. Hill
Family: Umbelliferae

Description and Distribution

Parsley, a native of Sardinia, wildly and extensively grown in the Mediterranean region and the USA, is a hardy, aromatic biennial, umbelliferous, much-branched green herb, sometimes lasting up to 4–5 years, producing a rosette of finely divided radical leaves in the first year and a flowering stalk up to 100 cm high in the second year. It has rich green compound leaves 2 or 3 pinnate. Flowers are yellow or yellowish green in compound umbels, fruit (commonly known as seed) is 2–3 cm long, crescent-shaped, conspicuously ridged, consisting of 2 mericarps. Leaves and seeds are used as spice. The color of the dried herb is green. Its aroma is pleasant, characteristic, fragrant and spicy, due to the volatile oil present in the plant.

There are two main types of horticultural Parsleys : (1) the one cultivated for the leaves (var. *crispum*) and (2) the other grown for its turnip-like roots (var. *redicusum* Danert) roots are cut after the fruits (seeds) are harvested. The roots are sliced longitudinally to facilitate drying. Only the former type is cultivated in India. In the latter case, roots are cut after the fruits (seeds)are harvested. The seeds are used for extraction of the Parsley oil of commerce. The aroma of the seeds is less than that of the leaves; 2–5 cuttings of leaves are possible from each planting before flowering. Within the leafy varieties, Parsley has the following three types of foliage (1)

K. P. Nair, *Herbal and Acidulant Tree Spices*,
https://doi.org/10.1007/978-3-031-04360-4_15

plain foliage (2) the double-curled leaf and (3) the moss-curled or triple-curled leaf. The fleshy-rooted Parsley has plain celery-like leaves.

Area, Production and Trade

Parsley grows wild or is cultivated almost all over the world. Though no authentic statistics on acreage and production of Parsley are available, it is known that the USA is the biggest producer of Parsley in the world. About 750–1000 tonnes of dehydrated Parsley are produced in the USA, which is mostly consumed within the country and it is one of the most popular herbs grown in the country. The demand for Parsley in the USA is very high and as the country cannot meet the internal demand through its internal production, it is imported from countries such as Mexico and Dominion Republic and, to some extent, from Canada, West Germany, Haiti, France, Hungary, Belgium, Italy, and, Spain.

Crop Management

Climate and Soil Parsley is a cold-weather crop, growing best in rich, moist soil, amenable to deep cultivation. In India, it grows better at higher altitudes.

Direct Sowing/Sowing in Nursery Beds

Sowing is done in March-May on the hills and in August–November in the plains. In European countries, where large-scale cultivation is practiced, seed rate is 6–8 kg/ha. Germination of the mericarps takes about 15 days or more to complete, which can be hastened by soaking seeds in water a few hours before sowing. On account of the slow rate of germination, and the tenderness of the newly sprouted seedlings, it is preferable to sow seeds in nursery beds and later transplant the seedlings in the main field when they ae 5–8 cm tall.

Transplanting of Seedlings

In small-scale cultivation, the seedlings may be planted at intervals of 15–20 cm in rows, 30–40 cm between the rows. Comparatively narrower spacing is provided for seedlings of the turnip-rooted types.

Manuring/Fertilizers Dressing with a nitrogenous fertilizer like urea induces rapid growth of the seedlings. No authentic data, however, are available on experimental findings.

Harvest The leaves are ready for harvest in about 3 months after transplanting. The outer larger leaves are periodically cut off for several weeks. Harvest period can be considerably prolonged by preventing the plants from flowering.

Yield In commercial plantations, one hectare yields about 2500 dozen bunches at each cutting, and, 2–5 cuttings before flowering are possible from each planting.

The roots of the turnip-rooted variety of Parsley are usually dug out after the fruits (seeds) have been harvested. They are frost-resistant, but, repeated freezing makes them pithy. The thick roots are cut longitudinally to facilitate drying. The fruits are collected for the extraction of the volatile oil of commerce. In France, a hectare of land is said to yield 800–1500 kg of fruits.

Post-harvest Technology

Composition The composition of the green leaves is given in the following Table 1.

The leaves, stems and fruits contain a glucoside "apiin", which on hydrolysis yields "apigenin", glucose and a sugar, apiose; a second glucoside, consisting of luteolin, glucose and apiose. The fruits and plants from Delhi yielded 2.2% of apiin, but, the second glucoside could not be detected.

Table 1 The composition of fresh leaves

Details	Content
Moisture	68%
Protein	5.9%
Fat	1.0%
Carbohydrates	19.7%
Fiber	1.8%
Mineral matter	3.2%
Calcium	390 mg
Phosphorus	200 mg/100 g
Iron	17.9 mg
Carotene (as vitamin A)	3200 IU/100 g (International Unit)
Thiamine	0.04 mg
Nicotinic acid	0.5 mg
Vitamin C (ascorbic acid)	281 mg/100 g
The vitamin A content can go as high as 8230 IU/100 g	
Riboflavin and Biotin are also present	

Processed Products

Dehydrated Parsley In a modern dehydration plant, the Parsley plant is cut by machine harvesters. The cut Parsley plants fall into field trailers, which haul them quickly to the factory. Here, they are unloaded, passed through a reel to remove dirt and field debris, and triple - washed. The Parsley plant is fed at a uniform rate into wide conveyor belt, where it is inspected and yellow leaves and other defects are sorted out. Dehydration is accomplished in a 3-stage continuous belt drier in a drying time of only 30 min. Parsley is neither blanched nor sulfited for drying. Final moisture content of the finished product is about 4%. The whole dried Parsley emerges onto a mechanical and air-separating system, where the stems are removed from the leaves. The leaves are sold as "flakes" or "granules"; leaves and stems are ground to powder. The entire process of Parsley dehydration, as described above, from field to final packages, is accomplished in less than 2 h. In to day's quick modern dehydration process, 8 parts of the fresh Parsley are reduced to just 1 part or its dehydration ratio is 8:1.

Volatile Oil All parts of the Parsley plant contain an essential oil, well known as "Oil of Parsley" which is responsible for the characteristic aroma and flavor of Parsley. The oil is recovered by steam-distillation and is used mainly to flavor food products. The oil obtained from the flowering – tops is of the finest quality, truly represents the odor of the leaves, but, yield is very low (0.06%) for commercial production. Commercial Parsley oil is distilled either from aerial parts of the herb, bearing immature fruits (herb oil yield is about 0.25%) or from the mature fruits (fruit oil yield up to 7%). The herb oil possesses a superior aroma and is more esteemed than the fruit oil. There is considerable difference in the physico-chemical characteristics of the herb oil and seed oil. The seed oil contains apiol (Parsley camphor) and α-pinene, with small amounts of myristicin, aldehydes, ketones and phenols. The herb oil is reported to contain "apiol", but, no detailed investigation appears to have been undertaken on this score. The apiol is used medicinally for the same purposes as the herb, but, is of doubtful therapeutic value. Commercial apiol is frequently adulterated with tri-o-cresyl phosphate, which may have severe toxic effects.

Fatty Oil The fruits yield about 20% greenish-colored fatty oil with a peculiar odor and disagreeable flavor. The oil has a high content of petroselinic acid (up to 76%). It can be used for a variety of industrial purposes, such as, making plastics, synthetic rubber, lubricating oil additives and protective coatings etc.

Uses

As Food Flavorant Fresh leaves mask even strong culinary odors and are commonly used for garnishing and seasoning of foods; they are eaten fresh, incorporated in salads, and used as an ingredient of soups, stews and sauces. They are also used in meat and poultry seasoning. The leaves are also employed to make a kind of tea which is considered to possess antiscorbutic properties, since it a is very rich source of Vitamin C (ascorbic acid).

The roots are used as a vegetable in soups. The dried leaves and roots are used as condiments, but, the use of fruits for this purpose has been contradicted by some authors. The fresh leaves are a good source of iron, calcium, carotene and Vitamin C (ascorbic acid), which can be up to 460 mg/100 g.

In Medicine The herb is a diuretic, carminative, ecbolic, emmenagouge and antipyretic; it has for long been in use for uterine troubles. The juice of the fresh leaves is used as an insecticide. Parsley causes skin rashes in some people, and this is attributed to the presence of furocoumarin, bergapten. Bruised leaves are applied to bites and stings of insects, and the mericarps are used to get rid of lice and skin parasites. Extracts and infusions of leaves and roots on subcutaneous administration in mice produced a depressing effect on the nervous system.

In a series of experiments conducted on hearts of frogs, cats and rabbits, it was noted that small doses (2 drops) of 1% aqueous infusion of leaves increased the amplitude and slowed down cardiac contraction; large doses (5–10 drops) of 10% of leaf infusion were followed by depression of cardiac activity. Intravenous administration of the infusion as well as the extract caused a fall in blood pressure and dilation of the blood vessels. Rabbits and guinea pigs when treated with increasing oral doses of apiole during gestation period, embryos abort and die after 26–28 days. Apiole is also known to cause poisoning and sometimes fatal lesions in liver and kidney. Hence, a cautious approach to the use of Parsley oil is called for.

Rosemary

Abstract At first, the chapter will discuss the description and distribution of the plant. Subsequently, area, production, trade, crop management, post harvest technology, processed products, and end uses will be discussed.

Keywords Description · Distribution · Processed products · End uses

Botanical name: *Rosmarinus officinals* Linn.
Family: Labiatae

Description and Distribution

Rosemary of commerce comprises dried leaves derived from an exotic, leafy, evergreen shrub *Rosmarinus officinalis* Linn. of up to 2 m high, cultivated in Indian gardens in cool places for its pleasantly fragrant leaves. The leaves are narrow, about 2–5 cm long, and resemble curved pine needles. The shrub bears a few bluish, white or violet flowers.

Rosemary grows wild and is also cultivated in Yugoslavia, Spain, Portugal, France, and other parts of Europe as well as in California, USA. It is a native of southern Europe and grows wild on dry rocky soils in the Mediterranean region. It has been suggested as suitable for cultivation in temperate Himalayas and Nilgiris Hills, in Tamilnadu, India, which have dry to moderately moist climate. Its lovely name "Rosemary" joins two Latin words meaning "Dew of the Sea", because it thrives best where fog rolls in from the sea, as is the case along with its native Mediterranean region.

The color of the dried herb is brownish green. Its leaves have a tea-like fragrance. The crushed Rosemary, however, has an agreeable and fragrant, spicy aroma with a camphoraceous note. The taste has fragrant, spicy, pungent, bitter and camphoraceous notes. Rosemary is marketed only in the whole form.

K. P. Nair, *Herbal and Acidulant Tree Spices*,
https://doi.org/10.1007/978-3-031-04360-4_16

Area Production and Trade

There is no authentic information on the above aspects in India concerning Rosemary cultivation.

Crop Management

Climate and Soil Rosemary cultivation has been suggested as suitable for cultivation in the temperate Himalayas and Nilgiri Hills with dry to moderately moist climate.

Propagation

Rosemary can be propagated by division, cuttings, layers, and seeds. Seeds are slow to germinate, but the cuttings root easily. Cuttings, 15 cm long, should be planted in a specially prepared bed of sandy soil to a depth of about 10 cm. Due to its preference for vegetative reproduction, not much of genetic variability is available. This genetic variability is essential for any crop improvement program. It is useful in the following:

1. Isolating better oil yielding lines
2. Screening and isolating lines for their adaptability to different climatic conditions

Raina et al. (1994) investigated the radio-sensitivity of shoot cuttings of Rosemary and concluded that 2D 50 dose of irradiation appears to be around 5 kr.

Planting For commercial plantations, seedlings or rooted cuttings are planted in the field 45-cm apart, in rows 120 cm apart.

Harvest Harvest commences from the second year after planting to any time up to August, after the full flowering. Flowering is in May-June. The plants must not be pruned heavily.

Yield No official figures are available, yet, on yield of the herbage/ha. However, Rosemary leaves yield 1–2% volatile oil (Oil of Rosemary), which is used in perfumery and pharmaceutical industries.

Post Harvest Technology

Drying and Packing As the leaves are spiky and thin, they are suitably cut into pieces. They are then dried, in shade, and subsequently crushed or powdered and packed in consumer packs.

Dried Rosemary leaves have the following composition (Table 1):

The dried Rosemary leaves, on steam distillation, yield 1–2% of a volatile oil (Oil of Rosemary), used in perfumery and pharmaceutical industries. Additionally, leaves contain saponin, tannin, ursolic acid, carnosic acid, amyrins and betulin and rosmarinic acid. A phenolic fraction possessing anti-oxidant properties has been isolated from the leaves and also from the oil.

Processes Products

Volatile Oil This is obtained by steam-distillation of leaves, flowering tops and twigs. The bulk of commercial oil comes from the plants growing wild in Spain; with smaller quantities from France, Dalmatian islands, Unisia and Morocco. Indian

Table 1 Composition of dried Rosemary leaves

Details	Content
Moisture	5.7%
Protein	4.5%
Fat	17.4%
Crude fiber	19%
Carbohydrates	47.4%
Ash	6.0%
Calcium	1.5%
Phosphorus	0.70%
Iron	0.03%
Sodium	0.004%
Potassium	1%
Vitamin A	175 IU (International Units)/100 g
Vitamin B	0.51 mg/100 g
Vitamin B_2	0.04
Niacin	1.0
Vitamin C (Ascorbic acid)	61.3 mg/100 g
Calorific value	440 calories/100 g

requirements are met by import. The characteristics of the oil vary with the source and also with the parts of the plant distilled.; the finest product being obtained from the dried leaves, freed of stalks. The oil is a pale yellow or almost a colorless liquid with the characteristic odor of the leaves and a warm camphoraceous taste. It has the following characteristics:

Specific gravity at 15 °C: 0.895–0.920
Optical rotation at 20 °C: −2 to +15
Refractive Index at 25 °C: 1.466–1.472
Solubility: 1 part in 10 pars of 80% alcohol.
The chief constituents of the oil are: pinene, camphene, cineol, camphor, borneol, and bornyl acetate

Adulteration The oil is sometimes adulterated with eucalyptus, sage and turpentine oils and their fractions.

Uses

As a Food Flavorant Fresh tender tops are used for garnishing and to flavor cold drinks, pickles, soups, and other foods. The leaves are employed as a condiment; dried and powdered, they are added to cooked meats, fish, poultry, soups, stews, sauces, dressing, preserves and jams. They are mixed with sage in pork and veal stuffings, and, sometimes are also added to biscuits.

Rosemary is a potent or strong herb; hence, it should be used with restraint to obtain the best results in the culinary art.

In Perfumery and Cosmetics

Oil of Rosemary is used mainly in cheap perfumery like scenting soaps, hair lotions and denaturing of alcohol. It is also employed in room sprays and inhalants. Superior grades of the oil are employed for blending in "*Eau de Cologne*" and to flavor meats, sausages, soups, table-sauces and other food products.

In Pharmaceutical Industry

The Oil of Rosemary is also used in the pharmaceutical industry, especially as an ingredient in rubefacient liniments. It is officially mandatory in the preparation of some of the pharmaceutical preparations. It is a mildly irritant oil and has been used as a carminative. Internally, the oil may be taken as a stimulant in doses of a few drops; a 5% tincture is used as a circulatory and cardiac stimulant. The oil is useful to stop headache and in tardy menstruation. It is diaphoretic and is employed with hot water in chills and colds. An emulsion prepared from the oil is used as a gargle for sore throat. The oil exhibits antibacterial protistocidal activities.

All parts of the plant are astringent and serve as a nervine tonic and an excellent stomachic. An infusion of the plant with borax is employed as a hair wash and is said to prevent premature baldness. The plant has been found useful in atonic dyspepsia. Tea made of Rosemary is used in Central America to cause temporary sterility in humans.

Flowering- tops and leaves have a camphor-like odor and taste and are considered carminative, diaphoretic, diuretic, aperient, emmenagogue, stimulant and stomachic. They are used for vapor baths for the relief of rheumatism, paralysis, and incipient catarrhs. Dried leaves are smoked to relieve asthma. A decoction of leaves is employed as an abortifacient. Pressed juice of the leaves possesses a strong antibacterial action against *Staphylococcus aureus*, *Escherichia coli* and *Bacillus subtilis*.

Flowers are an excellent source of honey, and they impart their characteristic flavor to the honey extracted from them.

Reference

Raina R, Johri AK, Srivastava LJ, Kumar S, Rana RC (1994) Radio sensitivity of shoot cuttings of rosemary. Indian Perfum 38(1):37–39

Ashwagandha

Abstract The chapter will primarily focus on the unique medicinal properties of this rare Indian plant in the Indian system of medicine, namely *Ayurveda.* It will also discuss the details of the collaborative research project on *Ashwagandha* on the control/treatment of covid 19 infection and recovery between The All India Institute of Ayurveda and the U.K's London School of Hygiene and Tropical Medicine

Keywords Medicinal property · COVID infection · Recovery · Placebo · International collaboration · Ayurveda

Botanical name: *Withania somnifera* (*Ashwagandha*)
Family: Solanaceae

Common Indian Name: Indian Winter Cherry

The Indian Ministry of *Ayush* (a *sanskrit* word meaning age, long life) has collaborated with the U.K's London School of Hygiene and Tropical Medicine (LSHTM) to conduct an investigation on Ashwagandha for promoting recovery from the global pandemic COVID -19 infection.

The All India Institute of Ayurveda (AIIA), an autonomous body under the Ministry of Ayush, and LSHTM signed a Memorandum of Understanding (MoU) to conduct clinical trials of Ashwagandha on 2000 persons (infected with COVID-19) in three U.K cities – Leicester, Birmingham and London (Southall and Wembley).

Ashwagandha (*Withania somnifera*), commonly known as "Indian winter cherry", is a traditional Indian herb that boosts energy, reduces stress and makes the

Ashwagandha and ***Manathakkali*** – two unique Indian herbal spices with extraordinary medicinal values.

K. P. Nair, *Herbal and Acidulant Tree Spices*,
https://doi.org/10.1007/978-3-031-04360-4_17

immune system stronger. It is an easily accessible, over-the-counter pharmaceutical supplement in the U.K and has a proven safety profile. The positive effects of Ashwagandha have been observed in long COVID-19 infection, which is a multi-system disease with no evidence of its effective treatment or management. It added that the successful completion of the trial could be a major breakthrough and could give scientific validity to India's traditional medicine, the legendary *Ayurveda*

While there have been several investigations on Ashwagandha, to understand its positive benefits on many human ailments, this is the first investigation to date, especially with a reputed foreign institution, where its efficacy in combating COVID-19 infection is being examined. The U.K. participants have been randomly selected. For three months, one group of 1000 participants will be administered the Ashwagandha tablets, while the second group of 1000 patients will be assigned a placebo, which is indistinguishable from Ashwagandha in appearance and taste. Both the patients and the doctors will be unaware of the group's treatment in a double-blind trial.

The patients would have to take 500 mg tablets twice a day. A monthly follow-up of self-reported quality of life, impairment to activities of day to day living, mental and physical health symptoms, use of supplements and adverse events, if any, would be carried out.

It took over 100 meetings spanning a-year-and-a half to finalize the project through diplomatic as well as regulatory channels before the MoU could be signed. The project has been approved by the Medicines and Healthcare Products Regulatory Agency (MHRA) and certified by the World Health Organization-Good Manufacturing Practices (WHO-GMP).

Reducing Anxiety

Of late, a number of randomized-placebo controlled trials of Ashwagandha in humans in India have been conducted which has demonstrated the efficacy of Ashwagandha in reducing anxiety and stress, improving muscle strength and reducing symptoms of fatigue in patients treated for chronic conditions of some ailments.

It is hoped that after the completion of the above detailed investigation, Ashwagandha will be a proven medicinal treatment for COVID-19 infected patients and be recognized by the scientific and medical fraternity the world over.

Manathakkali

Abstract The chapter will primarily focus on this common weed growing in Indian household backyard for its unique liver cancer controlling ingredient Uttroside B.

Keywords Indian weed · Liver cancer · Uttroside B

Botanical name: *Solanum nigrum*
Family: Solanaceae
Common Indian Name: Black nightshade, *sukkuti keerai*

Never underestimate the unassuming Manathakkali, growing in one's backyard, abundantly, for, it may have hidden therapeutic properties hitherto unknown.

The recent discovery of a research team of the Division of Cancer Research at the Rajiv Gandhi Center for Biotechnology (RGCB), in Thiruvanathapuram, Kerala State, India, has received many international patents, including from USA, Japan, Canada and South Korea. The work has been published in the nature group Journal Scientific Reports. The research team has identified a naturally occurring compound in the plant, Uttroside B, a saponin, found in the leaves of *Manathakkali,* which has liver protective properties.

Liver cancer, mainly Heptocellular Carcinoma (HCC), is the third leading cancer, worldwide. "Sorafenib and Regorafenib are the only USA FDA approved drugs currently available to treat HCC. But, they have only limited therapeutic value. Additionally, they have serious toxicity/side effects in humans, when used. The investigation conducted at RGCB reveals that Uttroside B. isolated from the leaves of Manathakkali is more potent than Sorafenib and does not have significant side effects and does not kill as many healthy cells as Sorafenib does, while attacking cancerous cells.

K. P. Nair, *Herbal and Acidulant Tree Spices*,
https://doi.org/10.1007/978-3-031-04360-4_18

Ayurvedic Use

Manathakkali has been used in Ayurveda for liver health. In Tamil Nadu it is used to treat stomach ulcers, piles, and mouth ulcers.

In collaboration with CSIR-NIIST research institute, Thiruvanathapuram, Kerala State, which has developed a novel method to isolate the Uttroside B. the research team at RGCB, is currently evaluating the efficacy of the compound in treating fatty liver, non alcoholic steatohepatitis and liver cancer caused by food toxins.

The RGCB has been collaborating with the Oklahoma Medical Research Foundation, USA, for the research, development, and clinical translation of Uttroside B. The compound is being manufactured by Chemveda, a pharmaceutical company based in Hyderabad, Andhra Pradesh, India.

The technology is now being developed in the USA-based biopharma company QBioMed, for which RGCB has received the licensing revenue. Recently, QBioMed's Uttroside B. program has received the "Orphan Drug" designation from the USA-FDA (Food and Drug Administration) against HCC. The "Orphan Drug" label would otherwise lack sufficient profit motive to attract corporate research budget and personnel, and, can now go on a fast track development and approval of the drug.

The above developments hold much promise not just for India, but, the entire world.

Research on: 'Ashwagandha' is an easily accessible, over-the-counter nutritional supplement in the U.K.

SEED
EXTRAORDINAIRE

Manathakkali (Black nightshade) has

Fennel
flowers

Fennel

Fennel produce

Fenugreek RMt 354
Bold seeds
Tolerant to downey & powdery mildew

fenugreek Narendra methi 2
Tolerant to Cercospora leaf spot and Downey mildew

Coriander produce

Coriander crop

Coriander Variety UD 475

Coriander plant

Coriander

Coriander

coriander produce

Fenugreek produce